Ceramic Art

Innovative Techniques

Ceramic Art

Ceramic Arts Handbook Series

Edited by Anderson Turner

The American Ceramic Society
600 N. Cleveland Ave., Suite 210
Westerville, Ohio 43082

www.CeramicArtsDaily.org

The American Ceramic Society
600 N. Cleveland Ave., Suite 210
Westerville, OH 43082

13 12 11 10 09 5 4 3 2 1

ISBN: 978-1-57498-299-2

Publisher: Charles Spahr, President, Ceramic Publications Company, a wholly owned subsidiary of The American Ceramic Society

Art Book Program Manager: Bill Jones

Editor: Anderson Turner

Graphic Design and Production: Melissa Bury, Bury Design, Westerville, Ohio

Cover Images: "Full Moon Canyon" by Elaine Parks; (top right) Porcelain vessel by Gary Holt; (bottom right) "Cool Bowls" by Emily Rossheim

Frontispiece: "Mum Leaves Basket" by Shuji Ikeda

Printed in China

Contents

Preface

Defining innovation is a lot like defining success. It's difficult, if not impossible, to generate a wholly unique approach to making. More often, innovation happens incrementally and in subtle ways. In general the innovator is only recognized after a lengthy time of testing that proves her or his skills as a maker. Further, like success, innovation is subjective. Because we who work in clay use a material that is literally as old as the hills, and humanity has been using clay for as long as it's been humanity, our innovations have been piling up for a long, long time.

Some of the most exciting pots to look at are ancient Japanese pottery that can be traced to the Jomon period, which dates from 10000BCE to 300BCE. They're made using basic tools, but are anything but basic and really prove that--at least in our world of clay--innovation can happen without computers, or "new" technology, rather it can come from an intense understanding of the materials one has at hand. Understanding your materials and their limits is always innovative.

The information contained in this book works more like a deciphering tool than a glimpse at something new. While some information may be fresh to the you, the reader, all of the information here has been put to the test and has some real world application. However, I would argue that there is still excitement and real innovation happening with each one of these artists. Perhaps most importantly it's through the research these artists have done and their willingness to share that helps you learn something interesting to inform your own work

Art is research and, just like any science, this book is an exciting glimpse at some of what today's artists are doing.

Anderson Turner

Charlie Tefft
Patience Is Still a Virtue

by Leigh Somerville

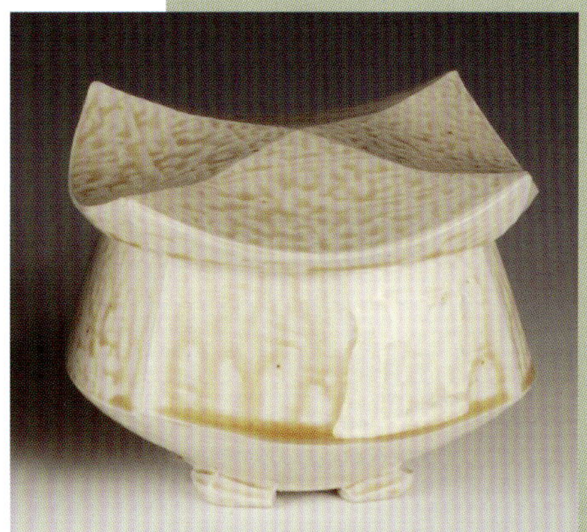

"Pagoda Jars," 4 inches in height, thrown and altered white stoneware, with added feet, sprayed with ash glaze and fired to cone 10 in reduction.

As Charlie Tefft strokes the belly of a recently finished piece, his gentle precision makes clear that the vessel has deep significance. While the claw-footed pitcher accurately models the Carolina Wren that hatched two sets of chicks while living in the artist's former studio, finding the shape took Tefft several tries. The finished product leaves no doubt that the brown speckles, rounded body and perky tail belong to the wrens that talked to him encouragingly each day while he worked.

Tefft is a patient man, and making art requires that skill. With the first wren pot, the tail wasn't perky enough, and it took some study to solve the puzzle: the angles had been cut too sharply. Tefft took the pot apart, recrafted it and now it sits saucily with others, ready to fly from their perches in the damp closet. In fact, Tefft says the crafting and recrafting process is one of the things he enjoys the most about what he does. He compares it to his love of playing with puzzles when he was a child.

"I love cutting up the pots and putting them back together again," he says. "I enjoy altering the pots. During this process, the pots take on a life of their own. As they do, I am able to find the ones that really work visually and physically."

Tefft explores certain themes and shapes in his pottery, and these continue to evolve. Often, as in the ancient Chinese and Korean pottery he admires, etched fish or slip-brushed grasses appear to move across the bottom of a series of bowls, each different, yet similar.

Rabbits, birds and fish have become prominent themes, suggesting movement and energy. "I am interested in the way they create or imply space within the pot, like they

"Night Rabbit," 14 inches in height, thrown white stoneware, with black stain and glass, sprayed with multiple ash glazes and fired to cone 10 in reduction.

are captured from a much larger space, or that they could take off and move beyond the surface," he said. Motion is a predominant thread in Tefft's work, and even the bottom of a teapot whirls like the spinning skirt that it models.

The son of a professional watercolor painter, Tefft discovered his own love of art growing up in Columbia, Maryland. He made his first bowl in kindergarten, fell in love with the first wheel he saw as a twelve year old, and learned to use one during a course at Goucher College.

Tefft says he benefited from the small classes in his Quaker high school and found that he had an artistic ability because of his dyslexia. He continued taking advantage of the Quaker educational system at Guilford College in Greensboro, North Carolina. He received his B.F.A. from Guilford in 1997, and began teaching pottery there past time two years later.

"Wren Pitcher," 9½ inches in height, thrown and altered white stoneware with black stain and oxide wash, sprayed with ash glaze and fired to cone 10 in reduction, by Charlie Tefft.

Tefft is among a very fortunate minority: He does what he loves, and he's making a living in the process. His position as a Guilford College lecturer, teaching others to work with clay provides a continuity that he finds invaluable. "It means that I don't have to re-orient my thoughts when I move from classroom to studio," he says. "I am always looking at pots, offering solutions to problems and seeing new solutions in my students' work."

Considering the small minority of graduates with fine arts degrees who are able to support themselves making art, Tefft is living every artist's dream. He says the emotional support of his wife, Danielle, and of his parents has helped make that dream possible. However, the responsibilities of being a husband, teacher and father have limited his studio time. "My decreased amount of time in the studio has helped focus my energy, resulting in more pots and more income from my art each year," he says.

Tefft enjoys the interaction with his students. As a teaching method, he transports work from his studio to the Guilford campus. There, in the campus studio, he glazes then fires the work in the gas kiln so students can observe those processes. Some of the pots are dipped in buckets of glaze, while most are decorated with brushwork images and patterns, then glaze is sprayed onto the surface.

While his students inspire him with their ideas and help him clarify his own, meeting the people who buy and use his functional art in their everyday lives is also part of the artistic process. He says he enjoys seeing his work in his clients' homes.

Tefft's professional experience began about ten years ago when he became part of a cooperative of artists in Atlanta and was able to take advantage of their gallery connections. When gallery owners came to the co-op to pick up other artists' work, they discovered Tefft's subtle earth colors and expressive yet functional forms. His attention to detail and line was unusual and dealers began to buy his work.

After the co-op shut down, Tefft found himself without a kiln. As luck would have it, he met a potter in Atlanta who needed help rebuilding her studio and learning how to use her new kiln. Tefft's experience with the same low-tech weed burner in college allowed him to barter his skills for the use of the kiln.

Tefft says the life of a young artist is easier when you can make what you need, salvage used and recycled materials and equipment, and ask for help. "I never felt like I had to have the best equipment, and I was able to find people who could help me when I needed help," he says.

In 2005, Tefft received a Freeman Grant and spent three weeks traveling through Japan with fellow faculty members from Guilford College. There, he was inspired by the architecture of the temples and shrines. He visited several potters whose work provided a connection to the ancient Asian art that inspires him.

When he returned home, he spent the first week feeling his way through the process of creating several 4-inch-tall pots. He named them pagoda jars, after the Japanese architecture that inspired them. Today, Tefft lovingly crafts his tiny jars, emulating what he first saw in Japan. "Their work was very accomplished, and I was struck by the amount of time they put into refining one piece," he says.

Cutting, Folding and Paddling

by Charlie Tefft

I throw the pitchers and pull the spouts before placing them in the damp closet to slowly dry. The damp closet dries the pots more evenly than setting them out in the studio to air dry. Once the surface of the clay is no longer tacky and the pot is still soft and malleable, I start the process of cutting, folding and paddling.

When cutting a **V**, I make sure the sides are equal lengths. This ensures that the lip will meet up once the top is folded together. After the seam is worked together, I use a metal rib to smooth the rough area so that the incision is hidden. Where the lip is joined together, there is a sharp angle that will tend to crack in the drying and firing. To reinforce the lip, I add clay and blend it into the lip. Once reinforced, I can start paddling to reshape the seam and soften the two pointed areas created by the fold. Now, I can shape the spout and pouring area. Once the reshaping of the body is done, I put the pitcher(s) back in the damp closet to stiffen up before I add the handle and cut the foot into a triangular shape.

After cutting the V shape (above), Tefft folds the lip together and gently works the seam. Later, he smooths the seam with a metal rib to hide the incision.

Takeshi Yasuda

Upside Down Porcelain

Some pots are thrown to the point of collapse, then inverted and stretched; they will remain inverted until they are dry enough to maintain their new shapes.

During the past few years, Takeshi Yasuda has given creamware—an invention of early 18th-century Staffordshire and long considered the preserve of industry—"a wholly new physical presence in the studio," according to British arts writer David Whiting. Yasuda discovered this traditionally lead-glazed, light-bodied earthenware by accident, but was immediately impressed by its "optimistic and visually liberating" appearance. For his creamware-inspired works he uses Limoges porcelain and higher temperature glazes "to achieve a mellow fluid liquidity."

Trained in Mashiko, Japan, Yasuda is known for his robust stoneware forms, "their complex profiles an elaboration of space and surface," observes Whiting. With the creamware, "there is that same spontaneous decision making, the interruptions of form made when a pot is taken still wet from the wheel"; however, the effect of working with a reduced repertoire, no color now, brought a "crisp and sharpened clarity" to his work.

His philosophy "is directly engaged with the notion...of the pot as a focus in our daily lives and rituals —not just a visual object, but something to be cherished on many other levels. While there are, of course, many other potters who share such concerns..., Yasuda is not a traditionalist in any conventional sense; he takes certain material from history and uses it for his own—often quite radical—deviations.

"He has an analytical mind, intrigued by the formal complexities of wheel-thrown pottery, its endless possibilities in terms of space, containment and enclosure, verticality and horizontality, and so on. But his concerns move beyond the purely conceptual. The energy he imparts in his work, its quality of stilled movement if you like, is much more than a 'style.' It is indicative of his wish, his need to engage and involve the user, and it underlines the openness of his conversation, both with the clay, and with us....He has been able to infuse his pots with that heightened sensuality denied to ceramics that merely function as utility objects."

At the same time, "Yasuda is far from being a puritan, and dislikes that connotation of functionalism that denies ornament and decoration. For him, these are special qualities in ceramics—indeed, as he would see it, part of their function. This function, in Yasuda's view, should not be discussed merely in terms of design and ergonomics, but in the way a pot can generate and be part of a ritual, add depth to life."

"Le Bol," approximately 6 inches in diameter, wheel-thrown and manipulated porcelain.

To express the "clay's dynamic," Yasuda throws tall porcelain cylinders on the wheel and allows them to collapse. They are then hung upside down, stretched and left that way until the shape is set.

"The result," says Whiting, "challenges our preconceptions of what makes a pot a pot. It is almost robbed of its age-old stability. A further limit of softness is achieved, but instead of melting or imploding, a new tensility is created.

"Yasuda's success comes from his real precision as a thrower, and it's this quality of attentiveness, of heightened and concentrated observation, both through the fingers as well as the eyes, which he brings to pottery and the world in which he lives."

"Tall Vase," approximately 20 inches in height, thrown and altered porcelain.

Squared Casseroles

by *Mike Baum*

Square and rectangular casseroles glazed and reduction fired to cone 10 in a gas kiln.

During the 30 years I've worked as a potter, my customers have always given me suggestions on what pots to make. Many years ago, someone asked me to make a rectangular open casserole suitable for baking lasagna, brownies, etc. The design I came up with is made with two thrown sections and is large enough in the wet stage (10–15% larger than the finished piece, depending on your clay body) so that when it comes out of the glaze firing, it is the right size to fit a lasagna noodle. The bottom slab is usually thrown the night before the top section is made so it can stiffen up. I try to time the drying process so that both pieces are the same consistency when attached together. The following technique can be used to make all kinds of differently shaped pots.

Throwing

Using a bat rather than the bare wheelhead, throw a flat slab for the bottom of the casserole. I use 5¼ pounds of clay to create a 16 inch di-

Throw a slab for the bottom of the casserole.

Throw a low wide cylinder and cut out the bottom.

Pull on opposite sides of the cylinder to create a rectangle.

PROCESS PHOTOS BY JUSTIN POOLE

Square up the sides using yardsticks or boards.

Trace the inside of the top section onto the base.

Cut away the excess clay from the thrown slab base.

ameter slab (figure 1). Remove the bat from the wheelhead and set the slab aside to dry.

Center 4¾ pounds of clay on another bat and throw the top section as a low wide cylinder, 14½ inches wide by 2¾ inches high. I like to have a thick, round rim at the top, which helps protect the finished pot from cracking and chipping. After the top is thrown, cut the bottom out using a wooden rib to shave away the excess clay, leaving a ½ inch lip (approximate) around the whole inside (figure 2). This bottom inside lip makes it possible to attach the top and bottom sections without using a coil.

Altering

After the top piece has stiffened a bit, wire underneath it and shape it into a rectangle. The clay should be slightly "tacky" at this point but firm enough so it doesn't slump when shaped. Hold your hands about nine inches apart, grasp the rim at the top with your fingertips and pull your hands gently away from each other (figure 3). Repeat the same on the opposite side. Next, pull the corners away from each other on the sides that haven't been shaped. Continue the pulling and shaping process until you have a basic rectangle.

While the top is still flexible, hold

Score then apply slip to the slab.

Align the top onto the slab and press down to attach.

Press the bottom lip of the top section onto the slab.

Create stitch lines, then blend the top and base together.

Cut away excess clay from the bottom using a metal rib.

Smooth the slab and wall transition using a rubber rib.

two rulers or cut yardsticks on opposite sides of the form and push all the sides in slightly (figure 4).

Assembly

When the top is leather hard, pick it up and place it on the bottom slab. Trace the inside (figure 5) and then cut around the outside with a fettling knife. Remove the cut pieces from the bat (figure 6).

Lift the top from the bottom slab. Using a fork, score and slip the area where the top was sitting and apply slip (figure 7).

Place the top back on the bottom and align the two sections (figure 8). Press the bottom lip of the top section onto the bottom slab. Smooth with a sponge and flexible rubber rib until they are seamlessly joined together (figure 9).

Pull the tines of a fork upward along the outside from the bottom slab into the top piece. The resulting lines will look like stitches all around the bottom seam. With your fingers, smooth the marks out and meld the two pieces together (figure 10). Keep the pot on the bat to stiffen up a bit.

Finishing

Place a bat over the top and flip the pot so its bottom is facing up. First with a metal then a stiff rubber rib, smooth out the roughness where the

Bevel the bottom edge using a vegetable peeler.

Attach pulled handles using water or slip.

Create a pattern and reinforce the handle attachment.

two sections were attached (figures 11 and 12). Run a vegetable peeler around the bottom edge to bevel it (figure 13). Smooth the beveled edges with a damp sponge. Flip the pot back over. Now you're ready to attach the handles.

I pull the handles and then bend them into horseshoe shapes. Whatever your final handle or lug design looks like, make sure they will not extend far from the profile of the finished piece, otherwise they will be prone to cracking due to heating and cooling (and therefore expanding and contracting) more quickly than the rest of the piece. Wet the handle sides that face the pot and press them firmly on (figure 14). Push the handle ends flat and pinch off the excess. Decorate with your fingertips or stamps (figure 15).

Move the pot onto a fresh, dry bat so that the bottom dries evenly with the top. Allow it to dry slowly before bisque firing and glaze firing.

Ray Bub
Reassembled Ring Teapots

by Paul Park

PHOTOS: JON BARBER, SUSAN NYKIEL

"Keel-Billed Toucan Reassembled Hollow-Ring Teapot," 15 inches in height, wheel-thrown, cut and assembled stoneware, fired to cone 5 in oxidation.

The teapot has fascinated ceramics artists in both the East and the West for the past 600 years. It is the queen of pottery shapes, a formal puzzle with limitless solutions. As a result, the challenge of making something new, a unique and compelling functional teapot, is a demanding one.

Ray Bub had been intrigued with the teapot format for some time before he took his Southern Vermont College class to the Bennington Museum to see the pottery collection. After that study trip, one student, Dylan Lawson, mentioned that he would like to make a ring vase similar to the 18th-century ring flask (made to fit around a man's forearm) in the museum's American folk pottery display. Bub showed him how to throw a hollow ring, then attached an oval base and a bottle neck. Afterward, the demo remained in the studio.

Soon, Bub started thinking about adding a spout, handle and lid to this traditional ring vase shape. The resulting teapot had an elegant and pleasing form, but was almost mute in its completeness. Bub's first instinct was to give it a voice by adding some of the hand-sculpted animal figures he'd been putting on boxes. So he added a spotted jaguar to the lid, and perched another on the inside surface of the upright ring. He then made several upright ring teapots with different animals.

Later, he made his first reassembled ring teapot. Again, there was nothing planned about the process: when the hollow ring was at the leather-hard stage, he had cut it apart with a bow saw, planning to reverse a couple of sections to create a zigzag profile to the upright ring. But the open ends of the cut-apart sections were unmatching trapezoid shapes that would not reassemble into a symmetrical closed form. Unhappy that he'd ruined the ring and wasted the time he'd spent on it, he decided to try to salvage his investment, and began rearranging the arc sections in different ways. Immediately, he was intrigued by the visual possibilities. He joined the

"Pink-Green Oval Cross-Section Reassembled Hollow-Ring Teapot," 14 inches in height, thrown and assembled stoneware, fired in oxidation to cone 5.

A ring is thrown by joining two walls at the top, trapping air inside.

When the ring is leather hard, it is inverted and trimmed.

Sections are then cut at various angles using a bow saw.

The section is closed with slabs, traced and cut to fit each end.

When reassembly is complete, clay spacers and supports are added for stability during drying, then the form is positioned on a thrown oval base.

A handle is pulled from a lug attached to a ring section, then a thrown spout is shaped and attached.

The last step is to make a finial for the lid; several are made and the one that most successfully enhances the design is attached.

"Silverleaf Bonsai Tree Reassembled Hollow-Ring Teapot," 11 inches in height, fired to cone 5 in oxidation.

arc sections together end to end and out of order, then put flat slabs on the two open ends. Then he added an oval base, a spout, a handle, a neck opening and a lid, embellished with a keel-billed toucan.

Then he made numerous reassembled-ring teapots, all decorated with animal shapes—Madagascar chameleons, African elephants, African giraffes, king penguins, highland gorillas, Pacific puffins, ring-tailed lemurs, North American mountain goats, etc. He made some sales, but after paying commissions, there was not a great return on the time invested, so he continued to earn most of his income by teaching classes in the studio, and making and selling "conventional" functional ware.

During this period, he also submitted slides of his animal figure teapots to several prestigious national and international craft fairs and exhibitions, but was not accepted to the ones he most wanted to participate in. As time went by, he grew more dissatisfied. Did the animal figures he had been attaching to the upright-ring and reassembled-ring teapots somehow cause jurors not to choose his work? Put another way, did the embellishment distract from the integrity of the design? When discussing this work with customers, he found them referring not to the

"Orange Five-Pointed Star Cross-Section Reassembled Hollow-Ring Teapot," 19 inches in length, wheel-thrown, sectioned and reassembled stoneware, glazed and fired to cone 5 in oxidation.

teapot composition itself, but to the animals, with such comments as, "I love the chameleons," or "Elephants are my favorite animal."

An artist frequently encodes ideas in some form to tell a story, but it was clear Bub's story wasn't coming through. He decided to put his trust in the intrinsic, undecorated eloquence of the abstract forms, and he started making his first reassembled ring teapot without animal figures. It was his "Pink Pentagon Cross-Section Teapot."

Soon he began experimenting with round, square, pentagonal, distorted, oval, trapezoidal and star shapes, cutting them into various-length arc sections and reassembling them into balanced compositions, positioning the assemblage on a thrown base, cutting out the lid, then adding spout, handle and finial.

The idea for these teapots came out of a chance visit to a museum, the chance interest of a single student, the desire to solve a purely technical problem, and a number of false starts and accidents.

The finished pieces can retain an unplanned quality that gives them tension and fluidity. Often the forms shift in firing, in ways that cannot be anticipated. Typically, Bub lives with the leather-hard reassembled ring composition for a while before

"Grasshopper Leaping Reassembled Hollow-Ring Teapot," 16 inches in height.

he chooses the location of the base, the spout, the finial—all the details that will solidify the piece in the viewer's eye.

In the case of his "Orange Five-Pointed Star Cross-Section Reassembled Hollow-Ring Teapot," he "had originally imagined this was going to be a tall piece, but when I was unwrapping it, I thought immediately of the Taiwanese ceramist Ah Leon and a teapot he had made in the shape of a long tree branch. Thinking of this Yixing-inspired teapot, I laid my reassembled ring composition on its side, then defined it by the location of the handle and spout. It needed no base, as it turned out. But to maintain functionality, I had to turn the spout upward."

Since Bub began working with this form, he has seen an earthenware ring vase with an oval base, made in Apulia in central Italy in 340 B.C. He has seen ring vase examples or pictures from the Japanese Kofun dynasty, the Chinese Tang, Song and Ming dynasties, ninth-century Moorish Spain, and twelfth-century Persia, as well as the colonial pilgrim flask in the Bennington Museum. Though geographically and chronologically widespread, the form is still a rarity, which makes it possible to deconstruct it and reinvent it for a long time and in myriad ways, without approaching anyone else's work. Bub believes there is boundless territory for him to explore with just this one vehicle. As for the next idea, he's preparing for another accident.

Multi-sided Forms

by Don Hall

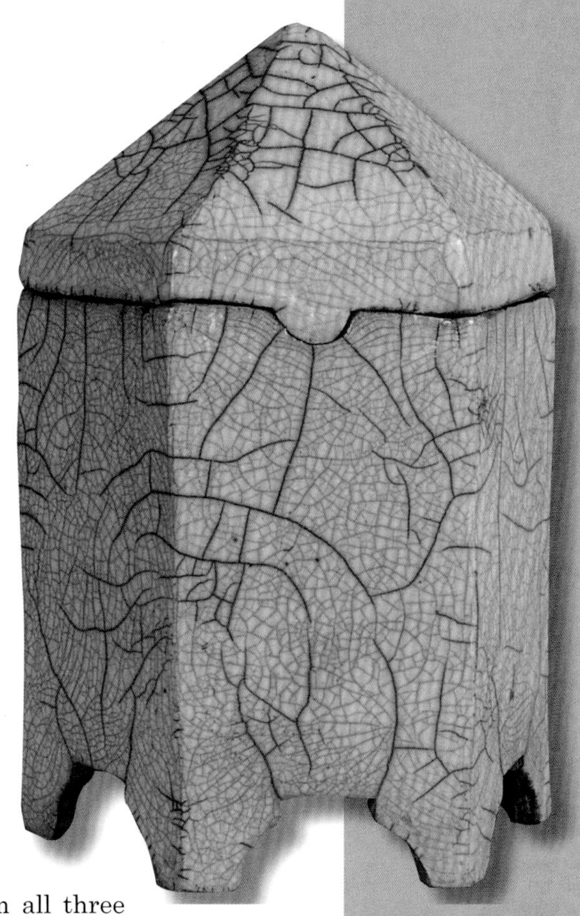

Hexagonal raku box, bisque fired to cone 04 and glaze fired to cone 07, post-firing reduction in newspaper.

Like many potters, I began learning pottery by throwing. After many years, I began handbuilding, purchased a slab roller and many contented hours followed. Here is a project on how to build a six-sided box with no throwing skills needed. The angles involved can be used for any six-sided form, so, by adjusting the measurements, you can make a piece of any height or width.

Make a template for the piece you're making. Include foot and lid pieces as needed. The box here will be 5 inches wide (figure 1). Roll out a ¼ to ⅝ inch thick slab and allow it to dry for a bit. Make a stencil from the pattern, mark the slab and cut out. Pieces should match (figure 2).

A six-sided form needs edges trimmed at a 30° angle. You can make wire cutter from a 2×3 inch piece of wood with a ¾×1½-inch notch (figure 3).

Dampen and score each edge. Fold up the sides and attach each one at a time to its neighbor (figure 4). The clay should be damp enough to not crack. Place coils on the inside of each seam and smooth them out (figure 5). Using a metal rib, clean up the outside of each seam (figure 6).

For the top, trim all three edges of the triangular panels to 30° (figure 7). Score and dampen the edges of each panel and assemble them (figure 8). Attach coils to the inside, smooth out then attach the lid to the base of the form (figure 9).

Now it is time to cut off the lid. Use a needle tool to score a line around the form. With a fettling knife held at an angle, cut off the top (figure 10). When cutting the lid, use a half circle in one side as a key so that it's easy place the lid (figure 11). Using the 30° tool, cut the edges off the sides of each foot segment. Assemble and attach the base adding coils to the seams (figure 12).

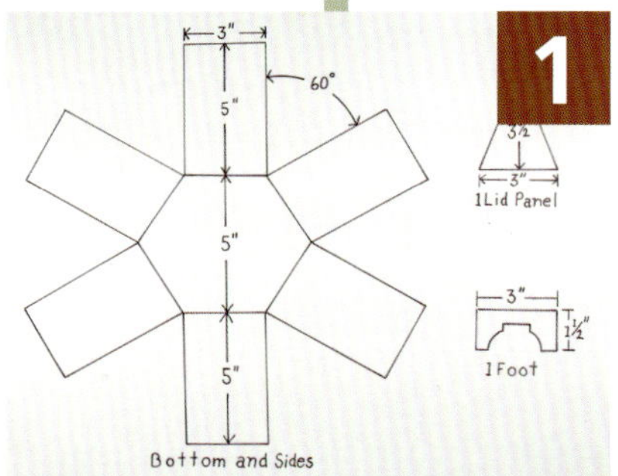

Make a template for the piece you're making.

Roll out a slab then make a stencil from the pattern.

You can make wire cutter from a piece of wood.

Fold up the sides and attach each oneto its neighbor.

Place coils on each seam.

Clean up the outside of each seam.

Trim the edges of the panels.

Assemble each panel.

Attach coils to the inside.

Cut off the lid.

Use a half circle as a key.

Attach the base.

Amy Santoferraro
Plate-O-Matic

by Paul Andrew Wandless

Amy Santoferraro combines just about any process, method or material with clay if it enables her to ultimately achieve the visual result her work requires. She's not alone in combining different methods and techniques with clay for new and more efficient ways to create work. It's more popular than ever these days to seek out different processes to use with clay and the work created is aesthetically exciting and fresh as a result.

One such combination of processes is screen printing directly onto clay, then using plaster molds with thick springy foam to press-form the clay slabs into a variety of shapes. This method is fairly quick to do from start to finish and has even earned the name "Plate-O-Matic" due to its ease of use and predictable reliability.

Ceramic artist Linda Casbon was giving a collectors workshop at Watershed Center for the Ceramic Arts and Amy was her assistant. Linda taught this method (which she learned from one of her students) to Amy during this time. Amy has since put her own unique twist on the basic process using screens. I always enjoy hearing these stories about how everyone learns from each other regardless of who is the teacher and who is the student! Clay folk are always happy to learn from anyone and then share with everyone.

You can create a shallow platter with a two-color, screen printed image using the following two-stage process. The first stage is to create the image by screening directly onto a prepped clay slab, and the second stage is to press-form the printed slab with a plaster hump mold into a thick, springy piece of foam.

Silk Screening Images

Amy transfers images onto clay using a silkscreen process. In this example, she begins with a rectangular-shaped slab of clay about $^{3}/_{8}$ inch thick, which she smooths with a rub-

Tools and supplies for this process include simple hump/drape molds and 3-inch blocks of soft foam. Texturing tools, cutters and printing supplies are all optional.

ber rib. The slab should be roughly 3 inches larger than the hump mold you're planning on using to assure it conforms to the whole shape. (Note: Amy uses terra cotta for this demo, but any clay body can be used.)

For a base color, she then coats the surface with porcelain slip (figure 1) brushed evenly across the entire surface with a wide brush. Once the slip dries a bit and the shine is gone, she smooths it with a rubber rib to remove any brush marks (figure 2). Although porcelain slip is used here, you can use any white or tinted slip—whatever background you want for your piece. Tip: Prep two or three slabs at a time so you have extras to work with.

Screen a Two-Color Image

To make a two-color print, Amy uses two screens with images burned into them using diazo photosensitive emulsion. Each screen is printed using a different color with the first screen being the background pattern and the second screen the primary image.

Commercial underglazes need to be the right consistency for silk screening to avoid bleeding edges on the image. To get underglazes to the consistency of honey, Amy leaves them open overnight so some of the water can evaporate.

Before printing on the clay, you need to load the open areas of the silk screen with color. Amy applies a bead of underglaze across the length of the screen (figure 3), then using a squeegee with a stiff rubber blade, she draws the underglaze across the screen into the open areas (figure 4). Next, she applies another bead of slip on the screen then carefully lowers the screen onto the clay slab

Prepare a slab and coat with slip.

Use a rib to smooth the surface.

Place a bead of slip on the silk screen.

Use a squeegee to charge the screen.

Place another bead of slip on the screen.

Carefully place the screen over the slab.

(figures 5–6). Once in place, she screens the image onto the slab creating a background of light blue circles (figure 7).

The second screen has several images in it so Amy uses wax paper on the bottom of the screen to block out all the images not being used (figure 8). The screen is then "loaded" with thickened black underglaze, lined up over the slab and screened over the blue circles (figure 9). The finished image is left to dry for 15–20 minutes or until it's dry to the touch. Once the image is dry, check the slab to see if it has stiffened enough to handle but is still flexible (figure 10).

Forming the Plate

Center the clay slab on a piece of thick springy foam and use a damp sponge to clean the surface of the plaster hump mold (figure 11). Be sure the piece of foam is larger than the mold being used. Place the mold over the area of the print that will be the final composition, taking into consideration how the shape and depth of the mold will interact

Squeegee slip to transfer the design to the slab.

Mask off areas of the screen you will not use.

A second screen with a second color is added.

Allow one slab to set up, but make sure it is still flexible.

Place slab on foam rubber and prepare a mold.

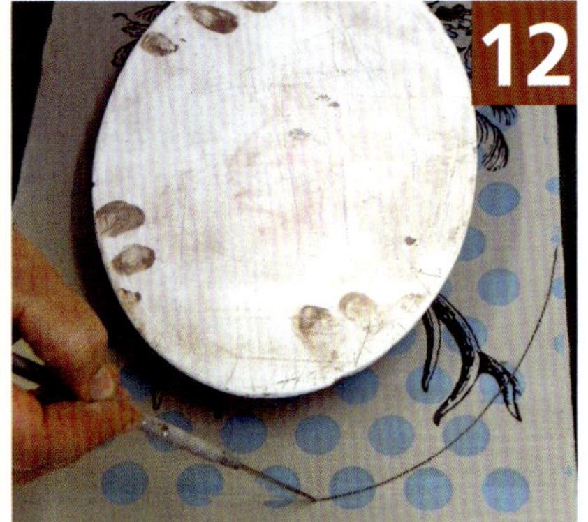

Trim excess clay from the slab before pressing.

with the image you created. Trim a wide border, leaving enough clay to conform to the mold (figure 12) and remove the excess slab. Keep your trimming tool handy because you'll need it after forming the plate.

Place both hands on the mold and press with slow even pressure until the back of the mold is roughly even with the surface of the foam (figure 13). While keeping pressure on the mold, trim and remove excess clay from the edge of the mold to create the rim (figure 14). Amy cuts at an angle so the rim also acts as a border. If you cut straight down, the rim will have more of an edge where the image or design would end at the perimeter.

Finishing Touches

To finish the plate, keep one hand on the mold, and flip the plate and remove the foam. Use a rubber rib to smooth the bottom of the plate (figure 15). Once the bottom is finished, flip the plate back over and remove the mold (figure 16). Finish the rim with a Surform tool and rubber rib.

Amy hand-glazed additional images on her plate. The finished piece

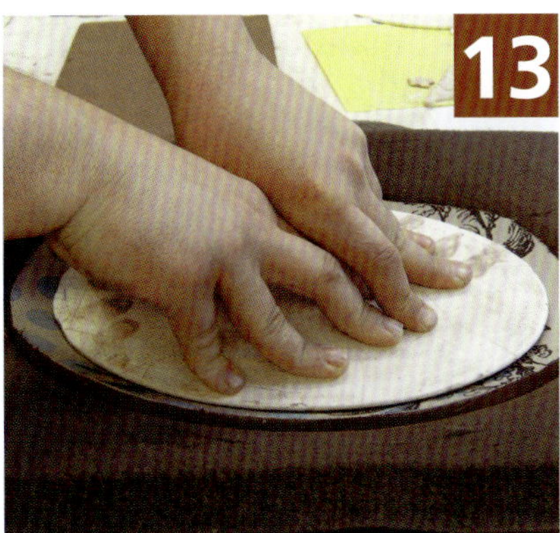

Use even pressure and press mold into clay.

Hold mold down and trim remaining excess.

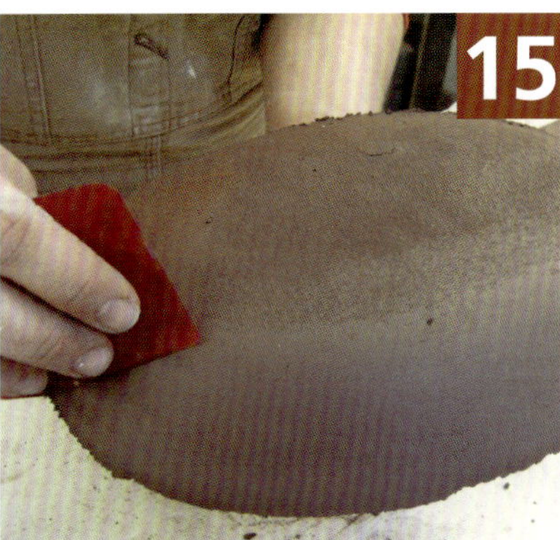

Keep slab on mold and use rib to smooth the surface.

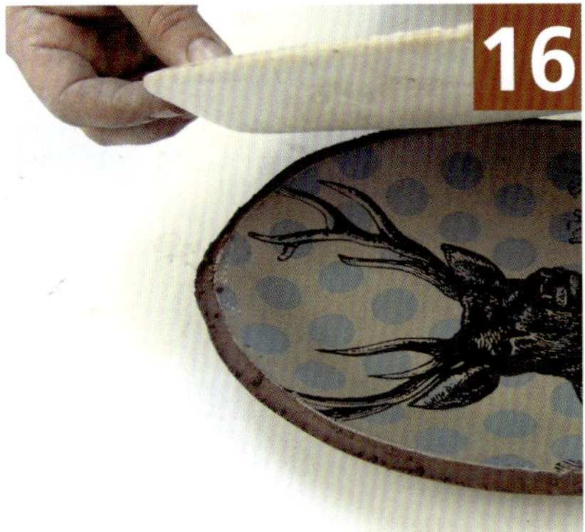

Place the completed piece on flat surface and remove mold.

As a variation, create a decorative edge prior to molding.

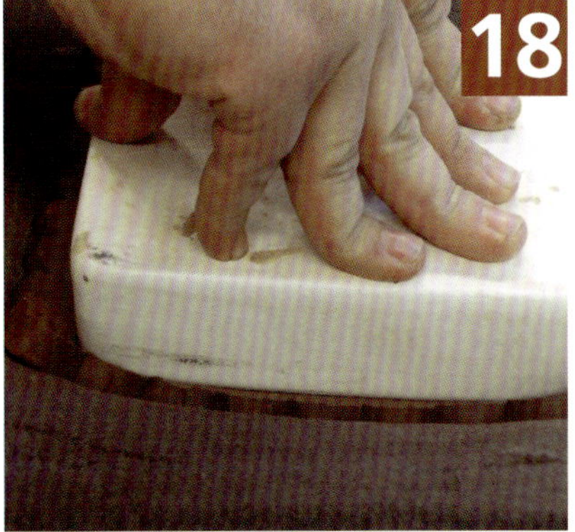

Press slab into foam.

looks wonderful and was simple to make. Once you have prepared slabs, this whole press forming process should only take about 15 minutes per plate.

Templates and Stamping

Like most techniques, you can vary this process. If screening isn't your thing, try one of these alternatives to make plates or bowls that are even quicker to perform and use common items.

Choose a template or form with an interesting profile or edge. This can be a plastic form or even a drawn shape that you designed yourself. Trace and carve the shape of its perimeter into the clay creating the edge of your plate (figure 17). Use stamps with interesting designs or patterns to emboss a design, pattern or composition into the clay; then line up the plaster mold and press into the foam creating the depth desired for the piece (figure 18). Once the form is pressed, remove and clean up with a rubber rib and other finishing tools as needed. This is a really simple way to create a plate with a complex embossed design

Piece showing scalloped edge and texture added before pressing.

Add feet to the bowls made with this process if desired.

(figure 19). Glaze or slip can be applied in the recessed areas (mishma technique) or a simple celadon can be applied.

For more variations, try using cookie cutters, pastry or tart pans or small dough cutters and cut out forms. Press the clay into the foam with a half-sphere plaster mold and add a foot if desired. Leave the mold inside to act as resistance to press against when making the foot. Amy uses a small coil and attaches and smooths it with her fingers (figure 20). A damp sponge can also be used to run around the foot for final smoothing of the surface.

These are just two variations that can be applied to this Plate-O-Matic method, and the possibilities really are endless. Just keep a few things in mind when experimenting with this process. Remember that the hump mold you're using must be slightly smaller than the slab so you're sure to get a good rim after pressing into the foam. The foam itself needs to be at least 4 inches thick and "springy" so you can achieve good depth in the plates or bowls. Seat cushions work well or you can get thick springy foam at a craft store.

The Making(s) of a Sphere

by Ursula Goebels-Ellis

"GEc3," from the "Industrial Spheres Series," 24 inches in diameter, wood-fired stoneware, with metal, glass and granite additions.

I have had my hands in clay for many years; however, what began as the hobby of a hausfrau has, over time, evolved into the work of an artist addressing philosophical ideas and global concerns. Today, my slab-built spherical sculpture is as much an expression of hope for world peace as it is metaphorical representation.

Materials

I work with a clay body modified from a Stephen Kemenyffy recipe: 30 parts Virginia Kyanite (35 mesh); 33 parts Cedar Heights Goldart; 33 parts Frederick Fireclay. Since plasticity is essential in my handbuilding process, I substitute mullite for kyanite and add 1½ to 2 parts paper pulp (dry weight). Toilet paper dissolves most readily into a soft pulp that can be mixed with dry ingredients or wedged into a commercial clay body. After adding the paper pulp, I allow the clay body to mature for a couple of days. I have kept this mixture for almost a year in plastic bags in an airtight container without excessive bacterial growth; however, it does not recycle well.

As I enjoy creating intuitively, I seldom sketch, yet there is intentional striving toward a particular form/design. Archaeological presence, technological advances and cosmic relations are key elements in communicating the desired message. While travelling, I collect indigenous materials from the depths of the oceans and the rims of volcanoes. At home, I work them into the clay, along with pieces of glass, scraps of metal and machine parts.

Process

I start each sphere by covering a concave mold with a piece of cloth. The mold can be a cracked bowl or a hemisphere made from plaster or recycled clay. I prefer the latter. The cloth prevents plaster chips from contaminating the clay body and assists in rotating the sphere without disturbing the surface; it also is useful in carrying the finished form to a place where it can be stored, glazed or fired.

A flattened piece of clay—the thickness varies from ½ to ¾ inch, depending on the size of the sphere—

Raku sphere from the "Celestial Sphere Series," 16 inches in diameter, by Ursula Goebels-Ellis.

is cut into a round shape for the base. Slabs are then attached to the base to build the sides. During the forming process, the edges are kept soft by covering them with small strips of plastic.

I use a fork to score the edges, apply slip and press the overlapping clay against the wall of the mold. For improved bonding, I use additional coils and always work the clay in the same direction (figure 1). When the walls reach higher than the mold, I insert a balloon for support (figure 2).

Super-sized balloons can be ordered through a local party or flower shop. For extra strength, it is sometimes helpful to insert one balloon into the other and inflate them only to two-thirds of their maximal size. With care, they can be reused several times.

Once the basic form is complete (figure 3), the drying process needs to be closely monitored. It is important to let air out of the balloon periodically to give the clay room to shrink. If the air pressure is too high, the drying clay cracks as it shrinks; if the balloon deflates too quickly, the whole structure could collapse.

While the clay is still leather hard, I paddle the sphere with the balloon inside to strengthen joints, or to alter shape and create additional texture. If one has not been left in the forming process, an opening big enough to insert a hand is carved out at this point. I remove the balloon and continue to work inside and out on form and texture, often adding found pieces of granite, glass and/or metal.

Special attention needs to be paid to how, when and where to incorporate such materials, because they dry, mature or melt at different rates and temperatures than the clay body. For example, to include a larger piece of rock or metal, I work like a jeweler putting a diamond into a setting of gold (shrinking clay). Volcanic rock keeps its form when raku fired but becomes a stream of lava in the much hotter anagama wood firing. Some metals run at mid-range temperatures, while heavier pieces of steel survive stoneware temperatures. Rocks can explode, while beautiful seashells dissolve into powder.

Once structure and design are completed, the opening is (partially) closed to form a small neck. The clay is then allowed to dry completely before firing.

Raku-fired Spheres

My "Celestial Spheres" are bisque fired to cone 08, then glazed and refired in a gas kiln. I usually limit myself to just a few glazes, such as Paul Soldner's Base White Crackle (80% Gerstley Borate and 20% Nepheline Syenite) covered with patinas, slightly overfired to 1900°–1950°F and quickly reduced. The glazed sphere is preheated and placed into the kiln; the firing takes about an hour or two, depending on size and structure, to the desired temperature. Removed from the kiln while glowing hot, the sphere is immediately placed into a metal container partly filled with combustible materials, such as leaves, paper or sawdust. The hot sphere ignites the combustibles and the container is quickly covered with a lid to prevent air from entering, and smoke from exiting.

I remove most of the raku-fired spheres from the kiln by gloved hand. Of course, I also wear protective heat-resistant and flame-retardant gear, including a face shield. This firing process not only challenges my physical strength and courage to take large pieces out of a red-hot kiln, but also the ability of the combined materials—clay, metals, rocks and glass—to survive thermal shock.

During construction of a sphere, the mold is lined with a cloth; it prevents plaster chip contamination of the clay and facilitates rotation without disturbing the surface.

A balloon inflated to match the interior space is used to support the top half of the sphere.

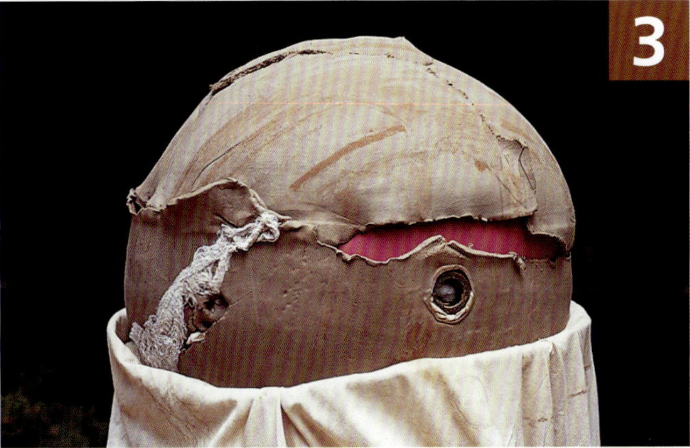

Additional slabs are then laid over the balloon to complete the basic form.

"GEy11," 18½ inches in diameter, handbuilt stoneware, with metal addition, wood fired.

Wood-fired Spheres

I seldom use glazes on my "Terrestrial Spheres." Rocks, metals, glass and texture convey global character, and the wood firing transforms and enlivens the surface. The bone-dry form is placed into the huge belly of the anagama, where flames dancing toward the chimney will create hues from deep brown to orange, and eventually stirred-up ashes will settle on the surface. A temperature of about 2400°F can be reached in two to three days; however, depending on the desired effect, firing time can range from a few days to weeks. For me, the long and somewhat arduous firing process becomes a wonderful spiritual/ritualistic ceremony of bonding with the elements that govern the universe.

Technological advancement has put humans literally and virtually into space. New knowledge of our place in the universe has changed communication and commerce, how we construe political and religious/philosophical concepts, and the way I make art. Through deliberate selection of indigenous materials, form and firing techniques, I am emphasizing the origin of all things, while the application of a spatial view captures a perception of global/universal interconnectedness and responsibility.

Shuji Ikeda
Weaving Clay

by *James Irwin*

As we stepped into the entryway of Shuji Ikeda's house in Berkeley, California, my wife and I were drawn to a large, handsome arrangement of irises with tiny white blossoms, a type of iris known in Japanese as shyaga. The assembly sprouted from a simple but dramatically flared vessel thrown from a rich, dark clay body.

The evidence of Shuji's dual passion for clay and flowers—and the context of his Japanese heritage—was visible everywhere. In one corner was a collection of vessels designed explicitly for flower arranging, or kado. Some were thrown; some were handbuilt. The variety of sizes and shapes seemed endless.

Shuji's kado pieces are highly sought after in the flower-arranging community both in the San Francisco Bay Area and in Japan. He views the line of kado pieces more as a vehicle for experimentation and play, however. His favorite form is the handbuilt basket. In front of the window was a basket display. These forms blend a solid architectural state-liness with an intricate tex-

"Mum Leaves Basket," 17 inches in height, slab and coil built, with Blue Rust Glaze, fired to cone 5.

ture of coils and strips of clay that have been wrapped, braided and woven, or fashioned into delicate twigs and leaves.

Process

Some of the baskets are glazed with a blue-green matt glaze, which Shuji calls Sei Shya (Blue Rust). Others are sprayed with iron or manganese oxides. Many have no surface treatment, but instead show off the dark, smoke-colored clay body from which they are constructed. In some, the dark body has been combined with a red clay by partial wedging, a traditional Japanese technique known as nerikomi. The degree to which the clays are wedged together results in varying effects when the clay is cut into strips or rolled into coils, then braided or wrapped.

I have been watching the evolution of Shuji's baskets for several years now. Earlier versions were direct interpretations of Japanese flower-arranging baskets known as hanakago, which are constructed from twigs, reeds or split bamboo. His newer work shows a more personal touch. "I am trying to create a kind of metaphor by mixing two ways of mimicking nature," he explains.

One way is to use the natural character of the clay—how it rolls, twists, breaks and bends. The other is the introduction of trompe l'oeil natural objects—twigs and leaves. Shuji calls this work tsuchi kago, literally "clay basket."

All parts are measured and cut from slabs and extrusions (figure 1). Joins are reinforced with extruded coils (figure 2). Once the walls are assembled, the top slab is attached (figure 3). Next, the legs are attached and reinforcement strips are added to the corners and bottom edges (figure 4). Notches are cut from the top for the extruded handle, and extruded coils are wrapped around the corners (figure 5). The coils are attached one by one (figure 6). The handle is wrapped with long coils (figure 7). Flat coils are carefully braided for side insets (figure 8). The braids are measured and cut to fit precisely (figure 9).

Learning Kado

The story of how these pieces came to be is the story of how a Japanese immigrant became a potter in America. The surprise twist is that Shuji's pottery teachers were not Japanese, but American. He arrived in the U.S. at the age of 23 (and likes to point out that he has now lived here more than 23 years).

He completed film studies at San Francisco State University, but because jobs in that field were scarce in northern California, he went to work selling Asian antiques for Sloan Miyasato at the Design Center in San Francisco. He was hired because he had translation skills, and was knowledgeable about Japanese pottery (which he had collected in Japan) and Asian antiques in general.

Sometime in the early '80s, he wandered into Pottery 7, the cooperative ceramics studio located in the Inner Sunset district of San Francisco, and signed up for lessons. His

PHOTOS: SHUJI IKEDA, RICHARD SARGENT

Woven ceramic basket, 17 inches in height, constructed from red and black clays, clear glazed.

All parts are measured and cut from slabs and extrusions.

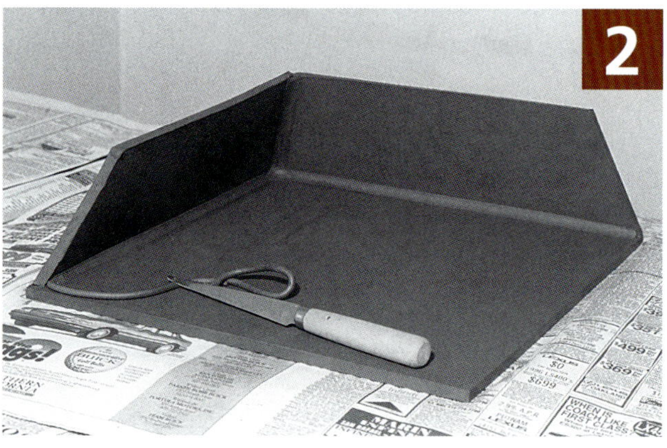

Joins are reinforced with extruded coils.

Once the walls are assembled, the top slab is attached.

Next, the legs are attached and reinforcement strips are added to the corners and bottom edges.

American teachers were somewhat amused to find themselves in the role of mentors to someone from the culture where ceramics is revered more than anywhere else in the world. Indeed, the work he began marketing through craft fairs five years later echoed classical Western shapes, and its only suggestion of Japanese influence was the raku firing.

Shuji traces his renewed interest in his native heritage to two events: moving into his own studio, and a decision to study kado. He explains that his kado teacher arrived in America in the early '50s and is thus more connected with older Japanese culture.

"My approach to learning kado was typically American," he says. "I told her to 'teach me everything in three sessions.' She laughed, gave me a bulb and told me to go home and plant it. Several months later, I cut the flower grown from the bulb and took it to class. She asked, 'Did you see how the flower broke the ground?' No. 'Did you see what the weather was like when it broke the ground?' No. 'Then how can you

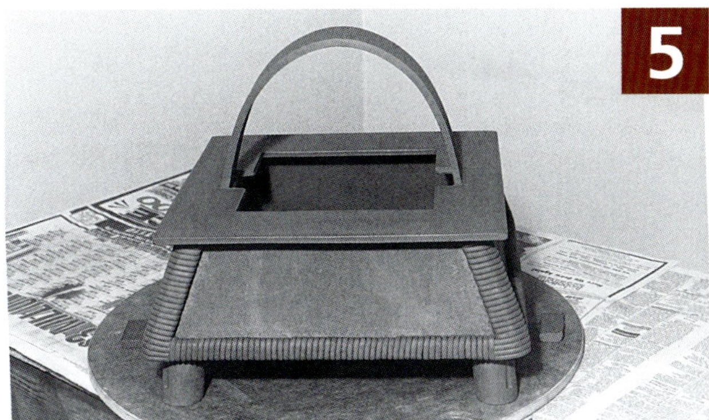

Notches are cut from the top for the extruded handle, and extruded coils are wrapped around the corners.

The coils are attached one by one.

The handle is wrapped with long coils.

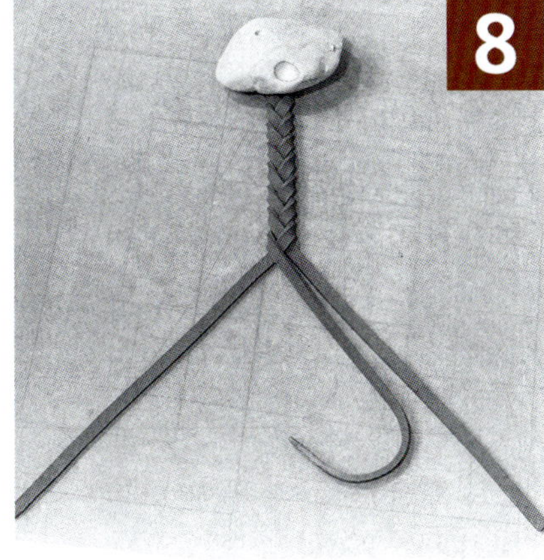

Flat coils are carefully braided for side insets.

The braids are measured and cut to fit precisely.

know how to arrange it?'"

Seven years later, he is still studying with the same teacher, and credits this experience with kindling a passion for "mimicking nature" in clay. He feels especially fortunate to have become immersed in two very different disciplines of artistic endeavor. "In Japan, if I had wanted to be a potter, I would have had to spend two or three years sweeping the floor. Sacrifice and hardship are valued highly. Discipline is traditional and external. There are rules. Manners and morality are important. In America, individual rights are important—you don't do anything you don't want to. Discipline is more internal, driven by the passion for what interests you. There is no 'way.' There are no rules. I struggled with that in pottery, always asking, 'What is the right way?' Until one teacher said, 'I don't care what you do or how you do it. Just make pieces that are beautiful to you.'"

Pointing to the shyaga arrangement, Shuji explained the functions of the various parts of the composition: "heaven," "earth" and "human"

Woven ceramic basket, 12 inches in height, clear-glazed black clay.

combined to express the theme that all creatures are supported by heaven and earth, which allows them to grow. He then showed me a book of "rules" just for irises; different types of irises have different rules, all the accumulation of a thousand-year-old tradition of flower arranging.

Later, he pulled out a photograph of a red flower seemingly tossed in a depression in white snow. "This throws out the rules," he said. "When you understand the rules, you can throw them away, but not until then."

In a way, both disciplines—Japanese and American—are different approaches to this same end, he remarked. "In my search for identity, I like to watch the interplay of bot

Recipes

Blue Rust Glaze

Cone 4–5

Barium Carbonate	32.5%
Dolomite	23.3
Custer Feldspar	37.2
EPK Kaolin	7.0
	100.0%
Add: Copper Carbonate	2.3%

Clear 3B Glaze

Cone 5

Gerstley Borate	47.5%
EPK Kaolin	24.2
Silica	28.3
	100.0%

Credit Card Dies

by Daryl Baird

A few simple tools are needed to create dies.

For several years, I had the opportunity to work alongside Jim Robison on the commercial exhibit floor at the annual National Council on Education for the Ceramic Arts (NCECA) conference. We worked the booth like a couple of traveling medicine men. First, Jim would draw the "townsfolk" in by demonstrating his considerable skill with a slab roller and an extruder, then I would sell them on the idea of personally owning one or both of these wonderful pieces of equipment.

I always enjoyed seeing how Jim could easily seize the attention of passers-by while he added beautiful touches to the vases and platters he built in just minutes, using simple tools he found in kitchen shops and paint stores. A pie crust ventilator made subtle scored lines in the clay while a tiny paint roller and a piece of lace trim gave it exquisite texture.

In addition to demonstrating at NCECA, Jim conducts workshops in Europe and North America. Among the most intriguing items he takes with him are the extruder dies he has fashioned from credit cards, membership cards and coffee cards.

Recently, a friend asked me to make an address sign for her new home. As a devoted "extrudist," I wanted to make the sign, at least partly, with my favorite studio tool. None of the stock dies I had on hand seemed suitable, so I decided to make the shapes I needed using Jim's credit card die construction techniques.

> **TIP**
>
> Clean the credit card with soap and water before starting to draw on it.

Making a Die

You'll need a few simple tools for making a credit card extruder die (above)—a no. 2 pencil, an indelible

Outline your design first in pencil, then use a marker.

Cut out the opening, working inside the line.

Trim burrs with an X-Acto knife.

Sand the die smooth.

marker with a fine point, a Dremel tool and assorted bits, an X-Acto knife and several No. 11 blades, emory cloth or 150-grit sandpaper, rubbing alcohol and a small rag. Safety glasses or goggles are essential when using the Dremel tool. Optional tools include a hand drill and bits, a scroll saw, a jeweler's saw and a small vise.

To begin, use a No. 2 pencil with a good eraser to lay out the shape of the die opening. Dull the finish of the card with fine sandpaper if the pencil marks are too light. Go over the pencil drawing with a fine-point, indelible marker, like a Sharpie (figure 1). If you make a mistake, these lines can be removed by wiping the card with a rag dipped in rubbing alcohol.

To cut out the shape you've drawn, a hand-held rotary tool, like the type made by Dremel, works fast (figure 2) and is fairly easy to control. These come with a variety of drill bits, along with grinding and sanding bits that can be used to refine the shape of the opening. In addition to using a Dremel tool, I tried cutting out the die shapes with a scroll saw. It worked well, but setting up the saw for this was tedious and time-consuming. I also gave a coping saw

TIP

As you clean up the die opening, hold the credit card die up to a bright light. This makes it easier to gauge where more trimming needs to be done.

Place the die on an extruder die to check the fit.

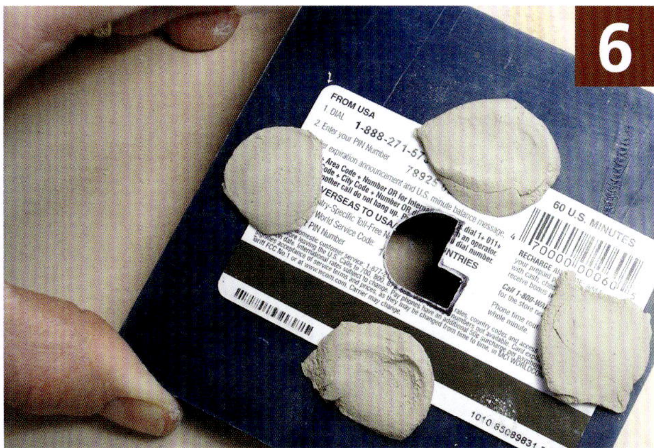
Attach the die with clay wads to a standard extruder die.

and a jeweler's saw a try. Even with a fine-toothed blade installed, the coping saw was next to impossible to use. The jeweler's saw cut more smoothly but it was slower than using a Dremel tool.

Once the opening has been cut out with the Dremel tool, it'll be rough, so you'll need to clean it up. I like using an X-Acto knife with a No. 11 blade to clean up the opening and square-up the corners (figure 3). The blades dull quickly when cutting plastic. Have several on hand and change blades often.

Once the die shape is done, use a small piece of sandpaper or emory cloth to clean off any burrs (figure 4). I used a narrow strip cut from a foam-backed sanding pad because it fits the contours and corners.

Using a Die

Credit card extruder dies are best suited for use with extruders that have a 3 to 4-inch barrel. Don't try to use credit card dies in large-barreled extruders because these types of extruders will exert too much pressure on the die, and cause it to crack. For the same reason, you'll achieve the best results by using only a two or three pound charge of very soft, well-wedged clay in the extruder to minimize the stress placed on the die. Place the card on an extruder die with a hole somewhat larger than the hole you just cut. I'm using a North Star standard extruder and the outer part of a small hollow die makes an ideal mounting plate for the credit card die. Looking from the underside, make sure the die is centered (figure 5).

Hold the die in place and turn it face up. Use small wads of soft clay to anchor the credit card to the mounting plate (figure 6). Then, attach the die to the extruder barrel and load the charge of clay into the barrel carefully so the credit card does not become misaligned.

Evaluate the first extrusion. If areas need to be refined, it's easy to go back and give the shape a little "tune up."

TIP

When working on large, flat projects, use a sheet of drywall as a work surface to minimize warping.

Street Address Project

To illustrate the use of these dies, I made a street address sign for my home. I rolled a large slab of clay that was just under a half-inch thick and cut out the oval shape using a plastic serving platter as a template.

So far, I have three credit card dies in my collection. I used the extrusion from my bull nose shaped die to create a decorative rim for the edge of the clay slab. The height of the notch in the extrusion matches the thickness of the clay slab. Make the bull nose extrusion long enough to cover the entire circumference of the slab and attach it as soon as it's extruded. Spray the slab before attaching the rim and use even pressure all the way around the piece to bond the trim to the slab. No scoring or slip is necessary.

I used a T-shaped die to make the numbers for the sign. The die is 1¼ inch wide by ¾ inch high. The "T" profile is easy to shape while also offering a large surface area on the underside, ensuring a strong bond between the extruded shape and the slab.

Make extrusions of several lengths and shape the numbers on a piece of drywall. Draw the outline of the numbers or letters you want directly onto the drywall, and follow these lines as you lay out the extrusions. If the numbers don't look quite right after the first attempts, you can go back and bend them more.

Keep the extrusions moist as you work. If you're assembling a shape from several extruded pieces, like the number 4 for example, take care to join the pieces thoroughly. This is where you'll need to score well and apply slip to the joints before attaching the parts.

Spray the oval slab with a mist of water and lightly place the numbers on the surface. When each one is properly positioned, repeatedly apply light, even pressure until the numbers are firmly in place. It isn't necessary to distort the shape of the extrusion to achieve good attachment.

Securely wrap the sign in plastic and allow it to "rest" on the drywall sheet for two or three days. Afterwards, slide the sign onto a fresh piece of drywall and lightly cover it with plastic. This will help it dry evenly.

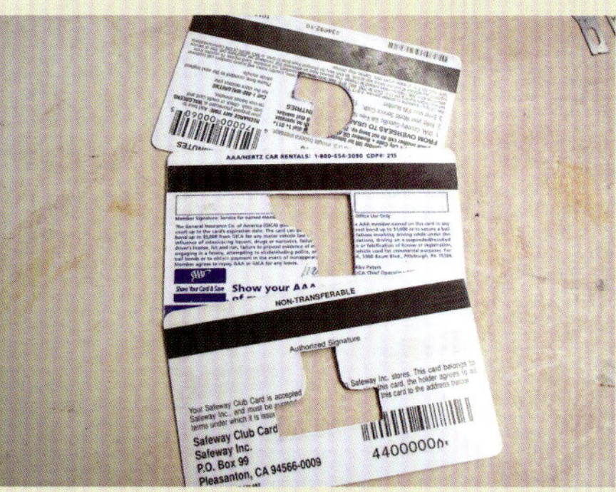

Three credit card dies.

Slab should match the thickness of the notch on the bull nose extrusion.

A T-shaped extruder die is used for the numbers, which are assembled on drywall.

Address sign, 18 inches long, Laguna Speckled Buff clay glazed with Laguna's Fern Mist, fired to cone 5.

The Printed Pot

by Mark Ganter, Duane Storti and Ben Utela

Two pots that were inspired by Southwest Native American pottery. These pots were printed using powdered slip and binder in a three-dimensional printer (notice the striations where each layer of clay was deposited on the printer bed) and were then fired.

In the Solheim Rapid Manufacturing Laboratory (located in the Mechanical Engineering Building at the University of Washington in Seattle), our research focuses on new and improved methods to describe complex shapes in a way that a computer can "understand" and to fabricate those shapes in ways that the computer can control (a.k.a., rapid prototyping). Three-dimensional printing (3DP) is our favorite method of rapid prototyping, because the required equipment is not outrageously expensive and you can use just about any material that can be obtained in powdered form. While our initial research aimed to address a biomedical application (digital fabrication of alumina dental implants), it was not long before discussions with a co-worker led to consideration of other kinds of ceramics. (The university setting is nice, because the head of your receiving department, like our own Ben Jones, just might turn out to have an M.F.A. and lots of good, challenging questions.) This article presents the basics of 3DP and everything you need to know to put together the materials for producing ceramic art objects on a 3D printer.

History

About ten years ago, we embarked on a project aimed at using a new type of geometric model to support the creation of some very interesting shapes involving lofts or variable section extrusions. This would be like starting an extrusion with one die and ending with another, with continuous connection between the two. Traditional commercial modelers include some lofting capabilities, but major changes in cross-section (e.g. changing the number of holes) can cause traditional modelers to break down. If the software system used to represent these shapes required unusual flexibility, then the manufacturing system to produce these shapes would need to be unusually flexible as well. Enter 3D printing.

3D printing was invented in Emanuel Sachs' lab at MIT and first became available in the early 1990s. One of the companies to license the MIT-Sachs technology was Z-Corpo-

Two beamlike objects that change cross-section along their central axis. These forms were printed in Redart TerraCotta. These examples are fired but not glazed.

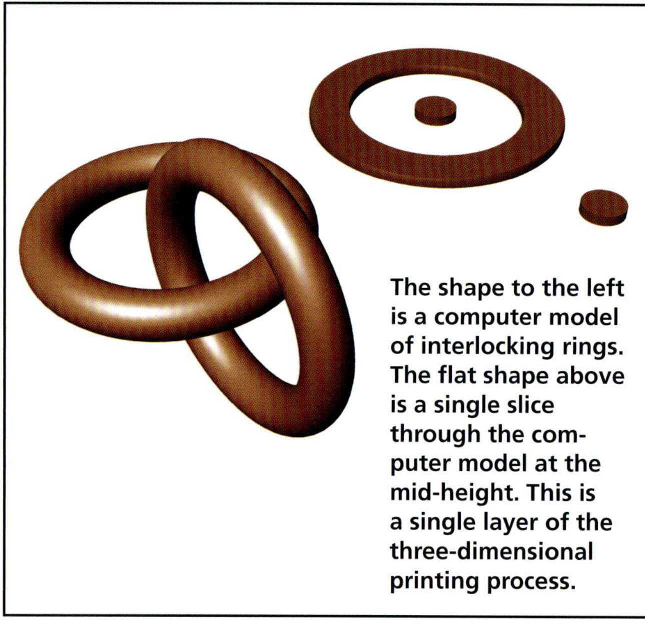

The shape to the left is a computer model of interlocking rings. The flat shape above is a single slice through the computer model at the mid-height. This is a single layer of the three-dimensional printing process.

ration, and they continue to produce 3D printers in a variety of sizes and capabilities. They can sometimes be found on internet auction sites for four-figure sums.

How 3D Printing Works

3DP takes a digital model and produces a real, three-dimensional object by adding powder in layers and selectively printing a binder on each layer that causes the powder to adhere in the area of the desired design. Typically, the digital model specifies a collection of triangular planes that wrap the surface of the object to be printed. A common file format is .STF which can be exported by most 3D computer-aided design (CAD) software packages.

The actual build process goes as follows. The 3DP system's software slices the object into layers ranging from 0.003 inches to 0.013 inches thick. A layer of powder matching the thickness of the digital layer (in our case 0.005 inches) is spread onto a build platform, or print bed. An inkjet printing system deposits binder into the powder layer corresponding to the image of the current layer. The print bed is lowered, another layer of powder is spread, another slice is printed, and the system continues until all layers are processed. When the 3D print is finished, our object composed of bound powder is supported in a bed of unbound powder. We now remove the unbound powder to reveal our finished object by a combination of manual brushing, vacuum removal and compressed

air. At times, one feels a bit like an archeologist at a dig site—and often with just as much excitement.

After an object is removed from the bed and de-powdered, one of a variety of post-processing techniques may be employed to "finish" the object, depending on its final use. Post-processing options include wax infiltration, epoxy infiltration, CA (CyanoAcrylate) glue infiltration, elastomer infiltration, or painting. These final steps often enable the part to function as a true prototype rather than just a form-and-feel object. For our ceramic-slip parts, post-processing consisted of kiln-firing, glazing, and glaze firing.

Adapting 3DP to Ceramics

In Sachs' lab at MIT, ceramics (i.e. alumina) were among the original materials explored for use in 3DP. Currently, however, suppliers do not provide art-ceramic powder as an option, which presented a large barrier to the use of 3DP by ceramic artists, so we decided to knock down that barrier. To do so, we would need to select a ceramic powder, decide on appropriate additives, and design the appropriate printing fluid; but what powder should we try first?

We recently had some students experimenting with casting metal into a low fire (cone 06) slip locally known as Xtra-White (from Seattle Pottery Supply). Since we had this slip on hand in powder form, it seemed like the obvious initial choice. We loaded the printer with Xtra-White slip powder, and used an existing alcohol-water solution as binder. It seemed like a good first test as simply mixing slip with water and letting it dry produc-

After objects are printed, the print bed is lifted and the unbound powder can be removed. In this case, several cups were printed at once. Inset: One of the cups sitting on test bars after firing.

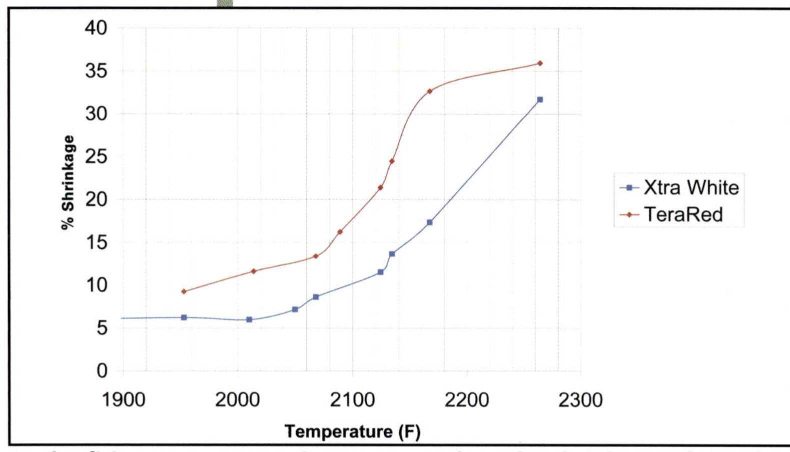

As the firing temperature increases, so does the shrinkage of the clay object. Because the objects printed with clay are very porous (like sponges), the shrinkage can be dramatic as the body vitrifies.

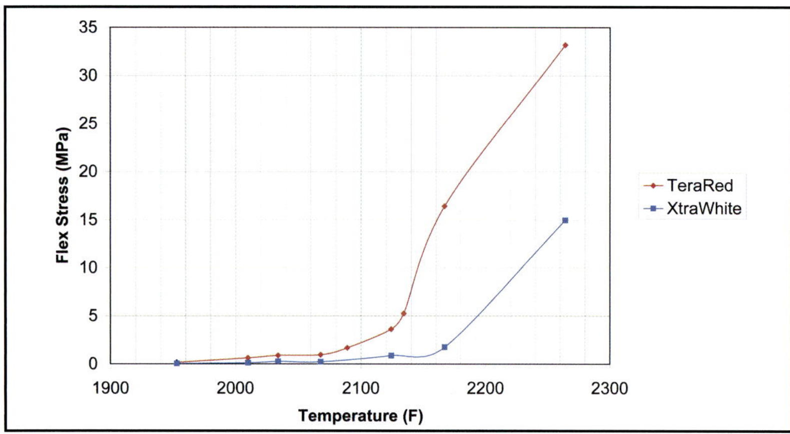

Flexural strength (measured in MPa, or Pascals) increases with higher firing temperatures. Because the clay body is so open, due to the printing process, strength does not increase appreciably until vitrification begins.

es a functional greenware. Let's just say our first tests were not terribly successful. The parts were so weak that any contact caused crumbling, and we could not remove the parts from the powder bed. However, the slip powder spread extremely well and had a very nice surface finish on the printing-bed surface.

We needed to find a water-soluble "glue" to add to the slip powder to give strength to the printed parts. The "glue" could also be added to the printing fluid, but keeping the glue in the powder bed avoids problems with clogged print heads. Again, the number of choices we have for in-powder "glues" is quite large, with PVA (polyvinyl alcohol), PVAc polyvinyl acetate), SCMC (sodium carbomethyl cellulose), PEO (polyethylene oxide) and various carbohydrates being high on the list of choices. Previously, we had run hundreds of test powder mixtures along with various water/alcohol binder setups. The process is not too different than glaze or clay formulation experiments (choose a test shape that is significant and keep good notes.) Based on experience, experimentation and a strong desire to produce a low-cost powder, we settled on a combination of PVA and extra-fine sugar (from the baking-supply aisle of the grocery store) as powder additives with the Xtra-White slip powder, along with an alcohol-water binder. After many more test runs (each one with different printing parameters), we finally succeeded in printing parts that could be removed from the 3D printer bed and depowdered. We focused on test bars that were 10×10×100 mm as they printed quickly and didn't require large quantities of powder to be mixed. Now, it was time to test fire the bars (in lots of five). Since the Xtra-White was a cone 06 slip, it seemed that a cone 06 firing (1828°F) was in order. When the test bars were examined after the firing, they crumbled to the touch and exhibited minimal strength. We continued firing more test bars (and gathering additional data) at increasing temperatures until the test bars melted to the kiln shelves. Having determined proper parameters for printing and firing, we were able to move on to fabricating simple functional shapes.

Encouraged by successfully printing with Xtra-White, we moved on to a higher firing slip, Redart TerraCotta. With the same powder additives and overall 3DP setup, we started printing terra cotta clay, and it worked immediately! We followed a similar testing process, firing bars in sets of five at progressively higher temperatures until the terra cotta slip melted to the kiln shelf. Again, we moved from printing bars to printing more interesting shapes.

Lastly, we arranged for engineering testing of about one hundred test bars to determine how the firing temperature affects both shrinkage and flexural strength. We thank our staff engineer Bill Kuykendall for his assistance. The results, presented graphically in the graphs on p. 38, allow for determination of a design point in the space of shrinkage, strength and firing temperature.

More Post-Processing Steps

With success and a bit more understanding of using the 3DP process with art-ceramic powders, we continued to explore more interesting object geometries. As the resulting fired objects are light and quite porous (they are essentially ceramic sponges), we discussed various ideas

These simple cups have been infiltrated with colloidal silica, fired, then glazed and fired again. The infiltration process makes the body denser and more durable.

for infiltration processes to reduce the porosity: infiltration with original base material slip, infiltration with colloidal suspension silica, and direct application of glaze. We chose to try infiltration with colloidal suspension silica followed by an application of glaze and then a glaze firing with quite good results. Lastly, we tried direct application of glaze. The results improved with either thicker glaze application or multiple coats.

Results and Conclusions

It is clear that 3DP can be used to create ceramic-art objects, out of three different types of slip bodies, and can be finished using standard ceramic equipment and processes.

We hope that at least some readers will be excited enough about the possibilities of this new approach to creating ceramic art to give 3DP a try. You should find that your 3DP objects readily accept both infiltration and glazes, and 3DP offers a variety of advantages to ceramic artists. You can print many copies of the same object; you can print many different objects at the same time; you can print interlocking/interconnected geometries; you can print objects in different sizes within a given print; you can print objects in different materials; and 3DP can provide increased access to ceramics for a broader practitioner base.

Our future work includes infiltration of the post-fired bodies with liquid slip (or possibly terra sigillata), continued adaptations of other slip bodies and additional engineering testing to determine how shrinkage, strength and porosity depend on firing temperature data. Finally, we wish to express our appreciation to the National Science Foundation for supporting the research that led us to ceramic 3DP. Now, find someone with a 3D printer and print something with clay!

Recipes

PVA Printing Slip

Ingredient	%
Xtra-White, Redart TerraCotta or Stoneware Buff Slip	62.50 %
Sugar (extra fine)	31.25
PVA (polyvinyl alcohol)	6.25
	100.00 %

Maltodextrin Printing Slip

Ingredient	%
Xtra-White, Redart TerraCotta or Stoneware Buff Slip	66.66 %
Sugar (extra fine)	16.67
MaltoDextrin	16.67
	100.00 %

The PVA Printing Slip mixture produced quite acceptable results (with all slips) but the PVA is a little more costly when compared to MaltoDextrin (which is available at the grocery store under the brand name Benefiber). The MaltoDextrin Printing Slip was also stronger in greenware form.

A solution of 16–18% ethanol (by volume) in water with a bit of food coloring works nicely as a binder. The food coloring lets you see if the fluid is deposited properly, and it burns out during firing. Our ethanol-water solution is vodka based (denatured alcohol tended to clog the printhead), so don't tell your graduate students how it's made.

Grace Nickel

Clay and Light

by Glen R. Brown

The combination of clay and light—a dense material and radiant energy—has perhaps held fundamental metaphorical significance since the earliest pit firings of prehistory. Matter and energy, two of the three fundamental components of the cosmos, seem condensed into pure expository form when clay meets flame. The ubiquitous terra-cotta oil lamps of the ancient world were in this sense ready paragons of a primary union, and it is no wonder that clay should have been analogous to flesh in stories of the origin of human beings. Combined with light, clay symbolizes animate matter, the weight of the body raised and warmed by the intangible but vital element of spirit. Because clay is extracted from the ground, the metaphor extends as well to plant life, with its roots in the soil, and to all the creatures that burrow into or walk upon the surfaces of the earth. The union of clay and light serves to represent life in its broadest sense: the precarious, temporary wedding of matter and a mysterious generative energy.

"Light Sconce #9," 18 inches in height, handbuilt white earthenware, with terra sigillata, vitreous slip and glaze, incandescent light.

The sculptures of Canadian ceramist Grace Nickel build consciously upon this ancient metaphor, joining hints of anthropomorphism and references to flora and fauna—especially insects—with actual light that emanates softly from the forms themselves. Although her works are in this respect ostensibly utilitarian—they function, after all, in the manner of lamps—their primary purposes are aesthetic and conceptual. They serve first and foremost to illuminate themselves. While they are capable of standing alone, Nickel's sculptures are ideally presented as parts of larger installations, with each component literally and figuratively shedding light upon the others. It is also in her installations that space, the remaining component of the cosmological triad, is most obvious. Her sculptures effectively utilize the passage of light across space to suggest a unifying energy in the natural world.

The final wall-mounted sculptures, essentially sconces, are made to fit over separately installed commercial light fixtures rather than to house the hardware and wiring itself. Nickel in fact, conceives of them as masks, shells or even exoskeletons—appropriate designations since they imply contingency upon some kind of living interior element. While a soft light rises from the open tops of these forms, the luminous focal points are the paired panes of glass that suggest glowing eyes. The intimations these create of a cognizant being are of course intentional, reinforcing the impressions of an inherent life energy within the material.

Knowing the primary inspiration for Nickel's work, one might expect the forms of her sculptures to be more thoroughly organic: characterized exclusively by flowing lines, rounded contours and a sense of malleability, without a straight edge or right angle in sight. The relationship of matter and energy, body and spirit, is not, however, the only fundamental dyad that Nickel has explored. Nearly as influential has been the connection between the organic and the constructed, the natural and the human made, which entered her work through reflection upon the relationship central to certain kinds of architectural ornament. When Nickel moved her studio to Winnipeg's historic Exchange District, she began drawing inspiration from the formal elements of the turn-of-the-century, neoclassical buildings there. Contemplating the volutes, wreaths, festoons and other ornamental forms, and recognizing the degree to which their stylization of natural elements balanced the shapes of the organic world with rational principles of design, she sought to establish a similar equilibrium in her sculptures.

Process

Handbuilding her forms in white earthenware, Nickel normally begins with large press-molded components that are subsequently altered and embellished through a process that could be likened to natural growth—the encrustation of ancient stone walls with lichens and mosses or the embrace of columns by creeping ivy. Working with a stiff paperclay slip, Nickel might employ a palette knife to articulate scale patterns like those on a green pinecone.

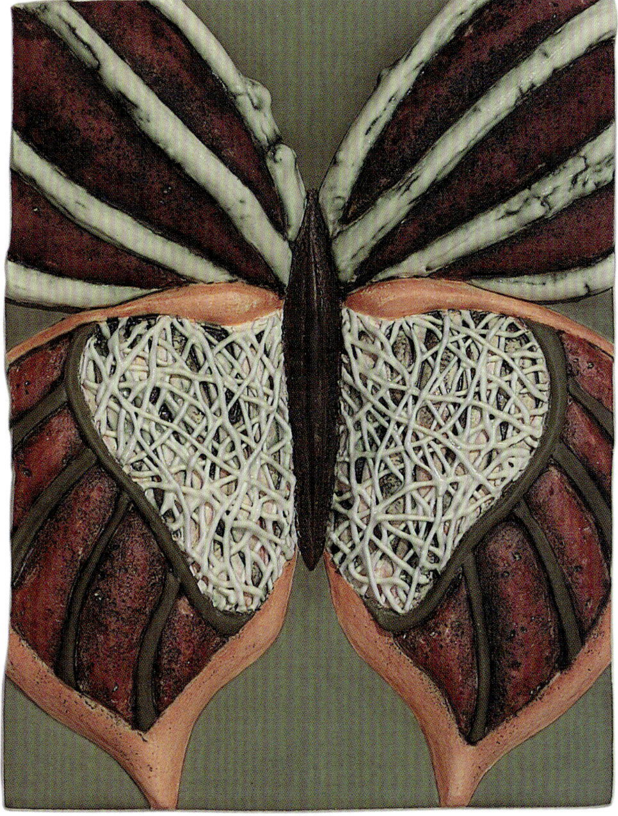

Tile studies, 7 inches in height, handbuilt white earthenware, with terra sigillata, vitreous slip and glaze. Her tiles are inspired by butterfly and moth motifs.

In other areas she might trail that same slip in a loose crosshatching that resembles interwoven layers of vines or meandering root formations wrapped tightly over constricted forms. Approximating the fibrous textures of rinds or striated plant stalks, she often impresses patterns into her surfaces with natural-object tools such as shells or fish bones. A V-shaped cross section of animal bone has, for example, proved ideal for producing textures resembling those of the softly indented surfaces of some succulent plants.

In order to accentuate these textures, both to heighten their impact on the eye from a distance and to reinforce certain rhetorical content having to do with light and shadow, Nickel permits the clay to dry and then applies a thin wash of black copper oxide and water to the entire surface, lightly sponging away all but the residue of this mixture that remains in the recessed areas. Color is added through the application of terra sigillattas, some in natural earthen hues and others tinted with various oxides, including cobalt and chrome. During firing the traces of black copper oxide burn through the opaque terra sigillatta, increasing tonal contrast and enhancing the visual effect of depth. The dark smudges of copper oxide also introduce a degree of organic irregularity, which for Nickel provides an important counter to the more-controlled aspects of the work.

The dynamic between control and accident, order and randomness, reinforces an underlying theme of complementariness to which Nickel has consciously adhered throughout the series. In addition to the pairing of matter and energy, with its metaphorical implications of body and spirit, and the dyad of nature and culture that is referenced in the combination of organic imagery and architectural form, a vaguer duality has often infiltrated her sculptures. Describing this as an ironic interdependence of the attractive and the repulsive, she attributes to it a curious influence over the creative process. Although admirers frequently describe her sculptures as beautiful, Nickel stresses that her inspiration sometimes derives from objects that she finds fascinating, yet at the same time, disconcerting or even sinister. The heads of insects, with their great unblinking eyes have, for example, influenced more than one of her works. In some cases, forms have derived from reflection on melancholic or even tragic themes. The process of altering and enhancing such forms, however, invariably softens them, leaving only hints of an unspecified pathos among the more aesthetically tranquil elements.

If these shadowy traces of pessimism had been omitted from Nickel's illuminated organic forms, her sculptures might reasonably have been described as decorative. Despite their deliberate cultivation of dualities, they might have proved incapable of moving the viewer on any level deeper than that of simple appreciation of formal dynamism. After all, their metaphorical evocations of a linked body and spirit only acquire poignancy from a vague apprehension of the frailty of this union. Life free from the shadow of death is in the end only a fantasy, and representations of this unreal state may, like pleasing patterns, momentarily charm the dreamer within, yet lack the power to spark a more lasting reflection. The dark spots of copper oxide in Nickel's works, suggestive of a creeping decay, couple with the ephemerality of the actual light emanating from the forms to confirm the artist's sensitivity to the imperative of the tragic in any art that purports to encompass the human condition. The pathos infiltrating the beauty of Nickel's sculptures is thus their guarantor of sustained relevancy. Ultimately, her works owe their success as metaphors to a willingness to embrace the inevitable and not merely the desirable in human experience.

"Light Sconce with Openwork #2," 14 inches in height, handbuilt white earthenware with terra sigillata and vitreous slips.

"Light Sconce #10," 19 inches in height, handbuilt white earthenware, with terra sigillata, vitreous slip, glaze, cast glass and incandescent light, by Grace Nickel.

Slumped Glass for Sconces

Nickel acquired some rudimentary glass-forming skills when she began her series of sculptures. The glass components, convex panes set into the ceramic frames from behind, were originally made through a simple process of slumping. After bisque firing the ceramic elements, Nickel cut flat sheets of float glass (ordinary window glass) into shapes corresponding to the apertures in the sculptures, leaving the dimensions slightly larger than those required in the final forms. Laying the sculptures face down in the kiln and setting the glass panes over the openings in the ceramic walls, she fired her pieces a second time to about 1472°F, a temperature considerably lower than that of the bisque firing. Slumping with the heat, the glass elements acquired the curving dimensions of the surrounding ceramic forms. At the proper moment, Nickel crash-cooled the kiln to about 1292°F, then allowed it to continue cooling naturally, fixing the glass elements with the desired curvature. Easily removed after, these elements can be sandblasted to give them a softer translucency and permanently reinstalled once the ceramic forms had passed through the final firing.

Nickel adopted the more complex procedure of casting the glass components in silica-and-plaster molds, a technique that not only allows for greater precision but permits her to incorporate patterns into the surfaces as well. The resulting glass forms are integrated into the ceramic structures exactly as before.

Phil Cornelius
Porcelain Thinware

by Judy Seckler

"Sea Change," 9 inches in height, press-molded porcelain thinware, sprayed with blue and green engobes, fired to cone 10.

One critic has called Philip Cornelius' work idiosyncratic and another labeled it impossibly thin, but those descriptions just scratch the surface of his evolution in ceramics.

His signature porcelain "thinware" developed back in 1970 from recycling curling wisps of leftover clay found on a bat after a piece had been wire cut, is often roughly textured and asymmetrical. The work is made up of elements that look like they've been retrieved from a lost civilization. The super-thin quality of his porcelain has to be admired for its alternating delicate yet strong skin. Cornelius has perfected this style into a reliable technique that has become his visual calling card.

"I like to do everything. I don't like to be cornered," says the artist. Of his process, he adds, "It changes like quicksand. You never know when you're going to fall through."

Cornelius is not interested in making conventionally pretty art. His aesthetic involves life on the edge, making sense of a debris-laden landscape like those portrayed in the films "Children of Men" and "The Terminator." A Cornelius teapot can be viewed as more tempest than tea. The idea of serene tableware has been exchanged for military equipment. The shapes of the pots look more like battleships, their spouts more like gun barrels with names such as "Patton," "Eisenhower" and "Sherman." The look, needless to say, is not for everybody.

Cornelius lives and works in a bungalow-style home on a shady, tree-lined street in Pasadena, California. The living room, with its bank of large windows, is sparely furnished but filled with the artwork from several artist friends. The dining room has an open tower of shelves in a

"Two Sirens Looking at the Tower," 9 inches in height, porcelain thinware, sprayed with blue and green engobes, fired to cone 10.

corner overflowing with work waiting to be shipped to various museum collections. A small office off the dining room has overhead bookshelves filled with many ceramics magazines and reference books. Beyond the tiny kitchen, a short hallway leads to his studio: a large, stark room filled mostly with tables, providing generous work surfaces.

He learned his method of press molding, or estampage, as a visiting artist in the town of Sevres, France, when the country's ministry of culture invited him in 1988 as part of its system of visiting artists. The pieces are the byproduct of his six-week European visit where he was exposed to the town's ceramics museum, which had a huge collection of 400-year-old molds. The technique has been an important part of his process since then. "It was quite an experience for me. I had never worked with molds before. I didn't know they'd be there," he says. He found that press molding gave him another visual language to add to his arsenal.

Thinware Process

To create his signature thinware, Cornelius starts with a huge slab of

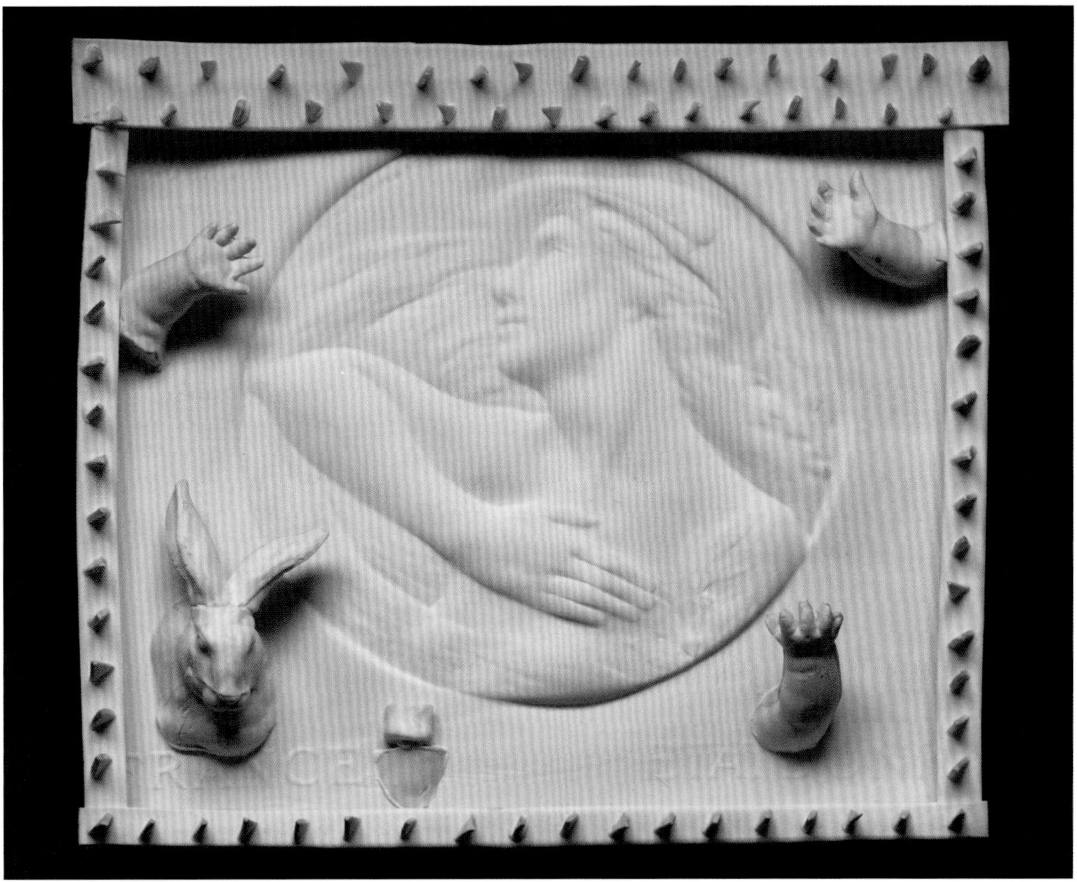

"International Rabbit," 9½ inches in height, Sevres porcelain thinware, by Phil Cornelius.

clay, 10 inches thick and between 100–150 pounds. He works on a plaster surface that's been cast from a piece of glass to create a smooth and flat surface. He lays the clay down on the plaster bat and runs a stainless steel wire through the bottom, leaving behind a section of clay that is one-twentieth of an inch thick. He builds his forms from these delicate sheets of porcelain.

As he discovered long ago, the forms are surprisingly durable after bisque and multiple glaze firings. Cornelius says that the right clay is key to the process. To build a piece, Cornelius lays out several sheets of porcelain from left to right on his worktable. The first pieces become the sides. The ends are joined and once the material firms up, he uses his breath to inflate the form so that it becomes hollow and is capable of standing up. Next, he assembles lids, rims, handles and spouts with the wetter clay.

Two-piece plaster molds are used to form baby heads, oranges, small birds and other details. The wet clay is introduced to the mold. Cornelius folds excess clay beyond the working surface. When slip is applied to the

"Four Birds and a Condom," 11 inches in height, porcelain thinware, sprayed with blue and green engobes, fired to cone 10.

Recipe

Engobe

Cone 10

Potash Feldspar	40 %
Kaolin	30
Silica	30
	100 %
Blue variation	
Cobalt Carbonate	5–7 %
Green variation	
Chromium Oxide	10–15 %
Orange variation	
Rutile	9–13 %

The blue, orange and green surfaces that resemble velvet are obtained by overspraying engobes onto the work.

excess clay, the matching pieces can be mashed together. A visible seam is formed around the piece. Sometimes, the seam becomes an integral part of the form and sometimes it's trimmed slightly.

In the last steps, the bottom and top of the sculpture are added. The bottom edge is dipped in slip and set on another sheet of clay where the bottom shape is trimmed to fit with an X-Acto knife. For the top, slip is brushed on to the top edge and a sheet of clay is added. Next, a thin sheet of plywood is placed on top and the vessel is turned upside down. The wood makes it easier to transport, as well as trim the top on a flat surface. Finally, design elements such as baby faces, birds, hands and decorative cylinders are attached.

Finishing

After a form dries, it is bisque fired to 1850°F. Next, Cornelius applies the gritty, textured engobe. He can apply 2–3 coats before the clay becomes too saturated. When it becomes too saturated, the form is placed in a kiln heated at 180°F for 15 minutes to dry out. Cornelius typically applies ten coats of engobe to a piece so that it has to be dried out in the kiln at least two and sometimes three times. Once all the layers have been built up, the work has to be handled with great care for its final firing to cone 10. Fingerprints can turn up in the final product, so Cornelius carefully lifts the work by the undersides.

Michael Wisner
Burnishing and Pitfiring

by Norbert Turek

PHOTOS: BRAD MILLER, MICHAEL WISNER

Burnished pots, two with polychrome slips and one with graphite, to 11 inches in height, by Michael Wisner.

Juan Quezada is the internationally known patriarch of a pottery revolution that has revitalized the village of Juan Mata Ortiz in northern Chihuahua, Mexico. In addition to his discoveries directly related to materials and techniques for building, burnishing, decorating and firing pots, Quezada's love of teaching has had such a significant impact that fully one-quarter of the village's residents are now producing decorated ceramics. Quezada has also taught numerous courses in the US. It was at such a course in California that Michael Wisner realized they "shared a passion for experimentation and discovery with clay and techniques."

Over the years, Wisner has adapted Quezada's forming and decorating methods to commercially available ceramics products, and subsequently has taught these modified methods. The students are often amazed at the simplicity with which high-gloss burnished pots are produced. Usually, the classes run only four to five days. Yet, by the end of a course, each student has produced an armful of coil-built, shiny black pots. In between classes, Wisner hunts the Elk Mountains for clay veins that might produce unique results—much as Quezada still wanders his surroundings in search of new materials.

Wisner isn't interested in sharing Southwestern pottery-making ideas

The bone-dry pot is carefully sanded, then covered with baby oil.

Evenly rubbed over the entire surface, the baby oil is allowed to soak in.

Rubbing in a circular motion with a damp cotton T-shirt fills small scratches and smooths irregularities.

After the surface is rubbed again with the damp cloth, the pot is attacked with a burnishing tool.

just to have more people making Southwestern pots, though. "Part of my passion and excitement is watching this art form evolve in the hands of sculptors, wheel throwers and handbuilders," he acknowledges. "Seeing how these rich surfaces may be used in the modern art arena is awesome."

He also experiments with techniques that veer away from his teacher's; for example, a gas-firing method that can more safely produce pots as richly black as traditional open bonfires; and a slip-casting body, developed with ceramist Richard Notkin, that will withstand the pressure of burnishing tools.

Finding the Right Body

"At first, I was frustrated because I would return to the U.S. (from Mata Ortiz) only to find commercial clays unsuitable for this style," says Wisner. But "Juan's passion for experimentation kept me going." After several years, he came across a commercial clay that worked—CT-3 from Mile Hi Ceramics in Denver, Colorado. A plastic talc/

For blackware, a graphite slurry is evenly applied with a 2½-inch-wide sponge brush to the entire surface .

After drying 1–2 minutes, gently rub the entire pot with a dry cotton cloth to remove the excess graphite.

For a mirror shine, burnish the surface in one direction, connecting every stroke, then burnish at a 90° angle to the original strokes.

ball clay body that achieves considerable strength in open-air bonfire firings between 1300°F and 1500°F, it "allows me to burnish to a highly reflective shine with moist clay right out of the bag."

While CT-3 clay works well, Wisner kept looking for a clay body that would fire at a lower temperature. He found that a 50:50 talc/ball body with 2–5% bentonite can yield excellent results. "I know the rule is 2%," says Wisner, "but 5% bentonite works fine, and adds considerable strength to the clay body." He calls this clay body recipe "Mike's Mud," and admits that it takes a little more work to produce than out-of-the-bag commercial clay, but the results are well worth the effort.

Wisner has tried many pot-forming techniques to see how each fares in the stressful bonfire process. Ironically, everything works—including handbuilt, wheel-thrown, slab-built and slip-cast pieces. Notkin believes slip-cast objects may actually have an advantage in that the clay has no "particle memory." They also have uniform wall thickness, which promotes even heating.

Preparing the Surface

"It is helpful for the artist to think of burnishing as preparing a canvas for painting. A well-prepared canvas will be smooth, allowing the paintbrush to glide uninterrupted over the surface," says Wisner. For years, he used clay slips as a way to achieve a smooth surface. After seeing Quezada's burnishing technique, Wisner gave up slips because of their limitations. "Slips leave evidence of their application whether done by brushing, dipping or spraying," he says. "They also sometimes flake off when polished or crack when dried. This is not to say that slips don't yield a beautiful surface. But if it's a smooth and shiny surface you're after, nothing beats a fine clay body that has the capacity to be burnished."

Materials for a bonfiring: metal bucket, sawdust, manure, firewood, vapor barrier, baling wire, lighter fluid and matches.

Once positioned, the oven-warmed pots are covered as quickly as possible with a metal bucket.

Preheated pots are carried from the oven to the firing site wrapped in cotton towels, then placed carefully on firing stands.

Sawdust (or dirt) is pushed around the base of the bucket to ensure an airtight seal.

The surface must be prepared for burnishing by a series of sandings at the bone-dry stage. First, Wisner uses 100-grit paper to remove any lumps and depressions left from handbuilding or the ridges left from wheel throwing. A little time invested in smoothing the pot while it's wet can save a lot of sanding time. Once the pot is evenly sanded with 100 grit, 220 grit is used to further refine the surface; 220-grit sandpaper erases the coarse tracks left by the 100-grit paper.

The most effective burnishing is achieved by moistening the bone-dry clay surface. Waiting for a pot to reach the right leather-hard state can be very tricky. A dry pot is virtually impossible to burnish, while a very wet pot that is burnished will lose its shine as it dries.

The story is told that everyone in Mata Ortiz used to work on the pots while they were leather hard. If a pot dried out, it had to be discarded. One day Juan Quezada's brother, Reynaldo, made a group of pots

Wood stacked around the bucket is held in place with baling wire.

Using lighter fluid on paper wads placed around the wood stack ensures an even burn.

After 20–25 minutes of burning, the coals are raked away to allow cool-down.

Examining results; sometimes, a propane torch is used to reoxidize (remove the carbon) from selected areas.

and left them covered with a cloth while he attended a nearby festival. Having perhaps enjoyed himself too much, Reynaldo didn't start work until late that next day. To his horror, he saw that the wind had uncovered his pots, and they were all bone dry. In an attempt to salvage his work, Reynaldo rubbed a mixture of oil and water on the bone-dry clay. To his surprise, it actually yielded a finer polishing surface than leather-hard clay. That happy accident completely changed the way pots were burnished in Mata Ortiz.

Michael Wisner's modification of Reynaldo's technique involves the use of baby oil and a damp cloth. A sanded pot is covered with baby oil and allowed to dry a minute or two until the oil soaks into the clay. Then, a soft cotton cloth (such as a T-shirt) is wetted and wrung out. The damp cloth is rubbed lightly in a circular motion on the clay surface. This removes small scratches in the clay and fills in small irregularities. Two or three wipes are usu-

To get traditional bonfiring results in a gas kiln, place 1–2 in. of sawdust or horse manure on a bed of sand.

A steel firing rack was constructed to maximize stacking space within a steel drum.

The drum is placed over the pottery and the sand is pushed against the bottom edge to ensure a good seal.

A cone 012 firing takes about 30 minutes, followed by a 30-minute cool-down before removal.

ally sufficient. Too much time on one spot or too much water, and the clay will become too wet.

Next, the pot is rubbed again with the damp T-shirt, this time going over the surface only once and being careful to work across the "rib" of the shirt (rubbing parallel to the rib of the fabric leaves visible lines). This eliminates the fine circular pattern left from the previous step, and leaves the pot very smooth and still slightly damp. The surface is now attacked with a burnishing tool.

Putting on a Shine

Wisner first makes all the burnishing strokes in one direction. After going over the entire piece, he returns and burnishes at a 90° angle to the first strokes. This enhances the shine and yields a flawless canvas that the paintbrush can effortlessly glide across.

His burnishing tool of choice is a 12-inch-long, stainless-steel automotive valve push-rod. "It polishes like no other tool I've ever come across," he remarks. "It lays down the surface completely smooth and regular with no small ripples, as you often see with a stone-burnished pot."

But all burnishing tools have their place—spoons, rocks, bones, chamois, even beans. Wisner recommends that all serious pot-shiners keep an array of them in their toolboxes. The diversity gives more options for stubborn clays or radical angles. Successful burnishing requires that the artist be aware of which tool is most effective at any given time. This can vary within minutes on the same pot.

Generally, Wisner does the initial burnish with the steel pushrod. Later, as the surface begins to dry, the steel rod can leave streaks. He may then switch to a polished deer bone or stone to finish the job. Another option is skin polishing, which leaves a satin finish, as seen on many oxidized (white- or red-colored) painted pots. "Put away the stones and roll up your sleeves," says Wisner. "I've found the soft skin on the underside of the forearm is a fantastic burnishing tool for satin finishes." Silk scarves or fine-mesh stockings are also fine tools for a satin burnish.

Slippery, but Not a Slip

Burnishing a thin coating of ground graphite is new to the American ceramics scene. This user-friendly technique yields a surface with a silvery, gunmetal finish—sometimes so metallic it's easy to mistake the piece for metal.

The graphite "slip" technique was developed in another part of the village of Mata Ortiz by a potter who noticed that when he signed his pots with a pencil, the signature remained metallic after the pot was fired. He began grinding pencil lead (graphite) into a powder and applying it to his pots. Quezada doesn't use this technique ("He thinks it's cheating," says Wisner), and no one in the village was willing to share their process with Wisner ("although you could see the graphite on everyone's hands"), so he experimented with various liquids to suspend the powdered graphite for application.

Water soaked into the pot too fast, leaving lumps of graphite on the surface. Oil allowed even applications of graphite, but later repelled the slip used to paint designs. Gasoline, diesel fuel and kerosene worked well, but were a little slow to dry. Then, during a workshop, a student knocked over the diesel and graphite jar. Desperate to continue the class, Wisner tried using some charcoal lighter fluid that was on hand for lighting the bonfires. It turned out that lighter fluid worked better than all the other media because it evenly spreads the graphite without repelling the design pigments. Furthermore, it dries faster, allowing the artist to begin painting almost immediately.

First, the pot surface is prepared for burnishing (as described earlier). For a matt-pewter look, the sanding and/or baby-oil treatment is skipped. Working in a well-ventilated area, Wisner applies the graphite mixture with disposable 2½-inch sponge brushes, that allow even applications with no evidence of brushstrokes. The brush is dipped into the graphite mixture and wrung out until very little liquid remains. This prevents overapplication and running of the graphite slurry. The surface is allowed to dry until no wet spots remain (normally 1–2 minutes).

At this point, he gently rubs the entire piece with a clean, dry, cotton T-shirt to remove some of the graphite. This step reduces the graphite layer to the point where painted slip designs can adhere to the clay body underneath. Wisner originally applied a thick layer of graphite, taking a more-is-better approach, only to discover that the hours (or days) of design work would flake off easily after the firing.

After the excess graphite is rubbed off (a microfilm is all that's needed), he burnishes as described previously. For sculptural work or on vessels where there is to be no painted slip work, he doesn't worry about removing the excess. At this point, the thickly applied graphite can be burnished with fingers, forearm skin or a chamois. He always burnishes the most visible parts of a piece first to ensure they have the best shine. On large pieces, he applies graphite to sections as needed to allow enough damp time for effective burnishing.

Hair Today, Brush Tomorrow

For painted designs, Wisner uses a thin slip made from Kentucky OM4 ball clay and, if painting on a white pot, mixes it with some black underglaze. The underglaze won't affect the final look, but the gray tone is easier to see on the slick surface. For fine lines, Wisner prefers a traditional Mata Ortiz brush made from human hair—a 1- to 3-inch piece of fine straight hair (15 to 20 strands is plenty) tied to the end of a stick (Wisner uses old bamboo chopsticks) with thread. Any "wild hairs" are removed to make a smooth brush that will pick up the clay paint without becoming too limp. Wisner says the women and children in Mata Ortiz are happy to provide a bit of their thick, straight hair. "You see them with locks missing all the time."

Ready, Set, Fire

Firing is extremely low tech and may be accomplished in a backyard, a fireplace, a gas kiln or even a barbecue. For the black metallic surfaces, an intense reduction fire is necessary. The process requires a nongalvanized metal bucket (galvanization burns off, producing fumes during firing that, in addition to being toxic, leave a fog on the pottery surface), such as a paint bucket, coffee can or oil drum. It should be fired once to remove any paints or residues that might affect the clay.

The pots are preheated in an oven, kiln or in the sun for several hours. This drives out residual moisture in the clay body and greatly reduces firing mishaps. It is critical to keep the oven or kiln temperature below boiling. At the boiling point, water in the clay wall expands from a liquid

Handbuilt jars, to 18 inches in height, and slip-cast fish vessel, decorated with human-hair brushes.

"3 Amigos," 13 inches in height, slip-cast native clays, by Michael Wisner.

to a gas (steam). The steam builds pressure rapidly and can easily pop out the pot wall.

Many fuels work in reduction fires. Thinly chopped firewood, cow dung and cottonwood tree bark all work well. Choose a site where the ground is dry. If this is not possible, a sheet of metal (not aluminum, which melts at 1000°F, or galvanized, for the reasons mentioned above) can serve as a vapor barrier. A 1- to 2-inch bed of sand serves as a base. This helps seal the metal bucket to create a reducing atmosphere. Well-dried dung (preferably range-fed animals) or sawdust from a pure wood source (not plywood or compressed board) is placed 1 to 3 inches deep on top of the sand.

It's best not to have the pottery touch the dung or sawdust because it sometimes causes blistering or undesirable flashing. (If flashing is desired, bisque fire the pot to cone 018 in a kiln and fire it on top of or partially buried in the dung.) Baker's parchment paper is used to separate the pots from the dung. Another option is to stack the pots on kiln furniture or metal grates. The clay will not fully reduce where they touch. If the opening of one pot is covered by another pot, Wisner puts a small handful of sawdust or dried dung in the lower pot to provide reduction for the upper pot's bottom.

The oven-heated pottery is transported to the firing site wrapped inside a cotton towel. Potters firing at a site far away from the oven can heat the pots (already thoroughly dried below the boiling point) to 300°F, wrap them in towels and place them in a cooler to keep them warm.

While a traditional bonfire is an easy, inexpensive method that yields excellent results, Wisner also achieves good results in a gas-fired kiln; not only is the firing easier to control, it is good for larger or thicker pieces (as well as in areas where bonfires might be illegal or inappropriate). This method was developed in response to the cold Rocky Mountain winters at Anderson Ranch (at more than 8000 feet elevation). The radical cool-down after bonfirings sometimes caused cracking, so Wisner began experimenting with the same firing set-up within a gas kiln. Doug Casebeer helped to refine the system using witness cones firing to cone 014.

Wisner still uses sawdust or manure, a steel base, sand and the steel can to cover the work (as described above) but, instead of firewood, a gas kiln provides the external heat. He built a small gas kiln specifically for this process. It fires to cone 012 in 30 minutes and cools within a half hour. Here, a cone pack with cones 018, 014 and 012 helps determine the stopping points.

For a traditional bonfire, newspaper, charcoal lighter fluid and a few coat hangers or other flexible uncoated wire are useful in the next step. The metal can is placed over the oven-heated pots, and dirt or sawdust pushed up around the outside edge to create an airlock within the container. Firewood is stacked 5 to 8 inches thick around and as high as the bucket. The entire stack is wrapped with baling wire to ensure that the wood will stay erect during the fire. Three or four wads of paper are placed evenly around the base of the wood, and doused with lighter fluid. The goal is to eliminate pottery breakage due to uneven heating.

The fire is lit and allowed to burn down completely. This will take 20–25 minutes. During this time, Wisner tends the fire with a long rake or stick, ensuring all the wood stays standing. A garden hose is kept nearby in case any peripheral vegetation catches fire.

Recipes

Mike's Mud

Cone 012

Texas Talc	50 %
Kentucky OM4 Ball Clay	50
	100 %
Add: Bentonite	2–5 %

In a clean bucket, slake the ingredients in a large quantity of water, then stir vigorously. Let settle 10 to 15 seconds. Draw off the top 2 to 3 inches of liquid "cream." Strain the creamy layer over a second bucket through fabric, such as a T-shirt, or a 100-mesh screen to remove organic debris and small pieces of grog. Add water to the original mixture in the first bucket and repeat until top water looks like skim milk after stirring and settling.

Let the strained clay (in the second bucket) sit for a few days—longer if clay is iron-bearing because the fine iron particles can stay suspended for several weeks. The mixture is ready to separate when the clay and water form two distinct layers. Decant the water. Stir the clay to rehomogenize. Pour/ladle it out onto a plaster board or canvas sheet. Dry to a workable consistency. Store in clean plastic bags.

Painting Slip

For decorating slip (painting), mix 1 heaping teaspoon of Calgon water softener with 1 gallon water; add 2 pounds Kentucky OM4 ball clay. Let settle for a few days. Draw off the water from the top and discard. Draw off the second "creamy" layer. Store in a jar; shake before use.

Note: Add a dash of commercial underglaze, such as Duncan Undercoat CC165. This colors the slip enough to see it when painting on burnished white clay but will not affect the color in the final fired body.

Graphite Slurry

To prepare graphite slurry, mix 2 tablespoons of fine graphite (available in tubes at most hardware stores) with 4 ounces lighter fluid (or substitute odorless lamp oil, diesel fuel or kerosene). Stir (or shake) it before use (in a well-ventilated area).

The black coloration of a fired piece is achieved by virtue of carbon impregnation. As the firing chamber heats up, the internal fuel (dung or sawdust) ignites, liberating carbon. For a special effect, Wisner uses a small propane torch to selectively reoxidize (i.e., remove the black carbon) areas on a reduced piece. This is best done immediately after the firing, while the pots are still hot. This technique can have dramatic results when using a low-fire red talc body, leaving nice red areas next to the rich reduced black.

Reduction firing is where the graphite really shines. Graphite stays on the pottery surface to a much hotter temperature in reduction (to cone 010) on most clay bodies. A safe window for this work is between cone 018 and cone 012. In that range, talc/ball clay bodies harden enough to have structural integrity and not dissolve in water. Additionally, within that range, the clay can maintain a burnish. Most lose their shine above cone 010. Therefore, to satisfy both key firing criteria, shoot for a range between cone 014 and 012.

The cool-down period lasts anywhere from 30 minutes to several hours, depending on the thickness of the pieces. To be conservative, pots may be cooled overnight and safely removed the following morning. Wisner works aggressively, cooling his pots only 30 minutes. This makes the firing about one hour from start to finish. Once the pots are cooled, they may be cleaned with a soft cotton cloth or washed with water. If all goes well, the fired pots will have an even black luster and the decorative designs, which will be matt black, will not flake or scratch off.

Form, Pattern and Smoke

by Jane Perryman

Double-walled vessel, 4 inches in height, coil-built T-material and porcelain mix, burnished, bisqued, then smoke fired.

Form can express a feeling or emotion. For many years, I've produced vessel forms that flare out from a narrow base, expressing a sense of uplifting of the spirits, of optimism, the kind of sensation one experiences upon reaching the top of a mountain. I have also been working with a form influenced by the traditional, wide, round-bottomed cooking vessel used throughout India. It's either made from clay for use on a domestic scale, or from beaten metal to feed large groups of temple worshippers. Although it balances on a tiny point, it is impossible to knock over, but will happily rock. Expressing qualities of security and of being well grounded, it is the antithesis of the flared forms, a metaphor for the opposing characteristics of the human psyche.

I focus on the vessel as a vehicle of expression. For me, the vessel represents the universal symbol for containing and offering, whether as nourishment for the physical body or spiritual soul. The essence of my work lies in combining the influences of traditional ethnic pottery and textiles with a contemporary interpretation. It is basically influenced—both in the making and firing processes, and to a limited extent, in the forms—by my experiences in India.

My first visit to India was to study Iyengar Yoga in Pune, a large city in the state of Maharashtra. While there, I discovered by chance a community of about 300 potters beside a busy dual roadway next to the river. These kumhars (the Hindu name for the potter caste) make a wide selection of work: thrown and beaten water pots, thrown cups, press-molded flowerpots, and coiled ovens.

In between Yoga classes, I returned many times to watch them, and was especially fascinated by a group of women coiling large tandoor ovens. I found great beauty in their economy of movement, where no action, however small, is wasted. I have heard this skill described as ancestral knowledge passed down from one generation to another. Many of the techniques and designs from the Harappan civilization (3000 B.C.) are still used in India today.

There are more potters (one million) in India than any other country in the world, but their work remains unknown. Perhaps it is because their status within the caste system is so low that their art is not recognized either within their own country or in the West.

At the time of my visit, I had been working with unglazed low-fired techniques for several years, drawing inspiration from early Celtic and African pottery seen in museums. It had not occurred to me that I would find a source of inspiration in urban India. That first visit was the beginning of a love affair with the subcontinent, culminating in extensive periods of research to collect material for a book on the potters of India.

I traveled from Tamil Nadu in the south to Himachal Pradesh in the Himalaya Mountains; from the western desert area of Kutch to the Gangian plains of West Bengal—a journey through rural India where life has remained unchanged for many generations. It was a profound experience, one that affected not only my claywork but also my philosophy and attitude toward life. To spend time in a country of developing technology where people struggle for the bare essentials of life certainly puts one's problems into perspective.

Although the demand for handmade utensils is diminishing in India, as factory-made crockery has become more available, numerous vessels, cooking stoves, furniture and architectural elements are still being made. An array of clay animal and human forms is also produced for religious worship amongst Hindu and Tribal devotees; this involves the concept of a wish fulfillment exchange with the god: "If you send rain for my crops, cure my wife of her illness, etc., I will give you a terra-cotta figure." After the initial ritual offering to the deity, the figure is left at the shrine to disintegrate, gradually returning to the earth. This ongoing cycle of creation and destruction represents the state of impermanence at the very heart of Hindu philosophy. In this way, a continuous symbiotic relationship exists between potter and devotee.

Process

My vessels are built by coiling, a traditional technique that is still used throughout India not only for pots but also for life-size figures—potters in the southern state of Tamil Nadu coil build horses and elephants up to about 16 feet high. The process of coiling requires few tools, and its versatility enables potters to produce any shape at any size.

In India, the clay is tempered with organic materials, such as rice husks or dry horse dung. In England, I work with a mixture of commercial clays—one part porcelain to two parts T-material (stoneware tempered with molochite); I cover the surfaces with a porcelain slip prior to burnishing. The technique is used throughout India to lessen porosity and develop a shine. To facilitate the burnishing, mustard oil is wiped over the slipped surface.

Coil-built vessels, to 10 inches in height, T-material and porcelain, burnished, bisque fired to 1760°F, then smoke fired.

Coil-built and burnished vessel, 10 inches in height, with resist patterns applied after bisque firing, then smoke fired in sawdust, paper and dung.

After a bisque firing to 1760°F, marks and patterns are developed with slip over paper and wax resists. Inspiration for this patterning comes from African carvings and textiles, especially the paintings on bark cloth by Mbuti women and the raffia pile textiles of the Shoowa tribe in Zaire, as seen in the ethnographic department of the British Museum in London.

Firing throughout India is carried out without kilns, in the open or in pits, using locally available fuels, such as agricultural and industrial waste (anything from rubber tires and wood to animal dung). In the Himalayan state of Himachal Pradesh, I saw potters firing a thousand pots in an open pile, using buffalo dung as fuel and reaching a temperature of around 1796°F. From this firing, only three pieces emerged damaged. Unfortunately, this kind of firing skill is disappearing, as low market demand compels potters' children to adopt other professions.

I have been exploring the aesthetics of low-firing techniques for many years now, and am smoke firing with

Smoke-fired vessel, 9½ inches in height, coil built, burnished, bisqued, decorated with slips over paper and wax resist.

Vessel, 8 inches in height, coil built, with resisted slip patterning, smoke fired, by Jane Perryman.

a combination of sawdust, paper and dung inside an oil drum or simple brick construction. I am also firing in saggars filled with sawdust and heated to about 1472°F inside a gas kiln; this affords a deeper black and a more even distribution of markings. I had played with the technique years ago, but never overcame the problem of saggar disintegration. Realizing I was spending considerable time making and repairing clay saggars, I abandoned the technique. Then, from Indian potters, I learned to replace them with metal saggars (discarded tin cans, for example); for their blackware, they use a sealed oil drum inside an oxidized firing.

The risks and unpredictability involved in smoke firing cannot be underestimated, but the rewards are immeasurable. I find it satisfying to be working with the opposite dynamics of slow control (coil building and burnishing) and fast, relatively uncontrolled firing, where I have an idea of the potential for surface marking but not the intensity of color.

Processes associated with low technology should not be equated with low skill, however. A gradual mastery of smoke firing is possible through experimentation and acceptance of failure as a tool for learning.

Animal Tracks

by Anne Macaire

For many years, I lived in the Yukon wilderness, where animal tracks were an important part of my world. Although we actually saw our wild neighbors only occasionally, the tracks they left behind told us of their comings and goings. They ranged from the cross stitching of a mouse on winter snow to the ambling of a grizzly bear along the beach.

One day, I found a particularly magnificent wolf track on the lake shore and cast it in plaster. I then began collecting tracks of all the other animals in the area; eventually, my collection grew to over two dozen species. Being a potter, I (of course) pressed these molds into clay. This led to a Canada Council grant and a project that spanned many years.

The same year that I received the grant, which allowed me to take a break from production work and explore the idea of track panels, we decided to move to town. Our relocation efforts extended over a period of five years and three towns, as we tried to find a place where our family would be comfortable. Through all of the upheaval, ideas for the animal-track project simmered on the back burner. I made 30 panels (each 1×2 feet) that moved around with us, were packed and unpacked, only to be thrown away when we finally settled.

Making large panels presented many challenges in developing a clay body, glazing, firing and mounting. Each was solved through trial and error. A rhythm of process evolved that allowed me the freedom to concentrate on design and glazes. In glazing the pieces, I kept to a palette that alluded to the natural world: earth tones, moss greens, lichens, stones. Rather than attempting a realistic backdrop to the tracks, I explored texture, pattern and color. This exploration took on a life of its own and led me down many paths I had not anticipated.

Paper Clay

To prepare a paper-clay body, I dumped the dry ingredients in a garbage can one-third full of water, just enough to make mixing by hand easy—stiffer than a slip but too wet to wedge. To this mix, I added paper pulp by volume. (I soaked shredded office paper for a few days or longer, then mixed it in small batches with an electric kitchen mixer until it was an unrecognizable pulp.) The paper pulp worked like magic. Large slabs cast from this clay body were strong, flexible, easy to dry with no warping or cracking. Note: paper pulp can also be used with a white clay body; the ink on the paper fires out

Recipe

Paper Clay Body

Ball Clay	9 %
Fireclay	13
Local Clay	65
Local Sand	13
	100 %

Add 3 parts paper pulp to 10 parts clay by volume.

Process

To produce a set of panels, I started with three plaster bats a few inches larger than the desired dimensions of the finished panels. A piece of old sheeting was laid on top of each bat and a simple frame made from 1×1-inch wood went on top of the sheet. The clay was then spread into the frame. At this point, I would decide what surface the piece would have. Some were smoothed flat with a long straight edge; others were stippled or sculpted in wave or folded patterns. For certain surfaces, it was

"Moose" (left) and "Moose Calf" (right), 5 feet high, paper clay with impressed plaster tracks.

"Wolf" (left) and "Fox" (right), 5 feet in height, paper clay, impressed with plaster tracks, fired to cone 06, by Ann Macaire.

more effective to wait until the clay was stiffer. Once the clay had set up enough to handle, I scored the edges around the frame so that I could remove it, and slid the slab, along with the cloth, onto a piece of plasterboard. I then pressed the plaster molds of the animal tracks into the clay.

When dry, the panels were fired on edge, with firebrick to support them, in an electric kiln. I also used firebrick to support the top shelf, which enabled me to fire six panels at a time. By firing to cone 06, I could do bisque and glaze pieces at the same time. Some panels were single fired; some only bisqued. Many were fired six times, but loss due to cracking was minimal.

To mount the panels, I used clear silicone adhesive (which can be found in any hardware store) to attach ½-inch plywood to their backs. Silicone adhesive contents and application recommendations vary widely. I spread it evenly over the plywood, then beaded it onto the clay panel. This may be more than was needed, but I wanted to feel confident the mounting would hold.

To apply adequate pressure while the glue dried, I laid the panels face down on a 2-inch-thick foam, then positioned three heavy firebricks on the plywood (attempting to use clamps resulted in breakage). The foam allowed even very delicate surfaces to be glued this way. The plywood, which was recessed 3 inches on all sides, allowed the panel to be hung with two "D" rings.

Linhong Li
Slab Paintings

by Yuqian Chen

Onglaze painting, approximately 24 inches square, with brushed pigments on white glaze ground.

The life, soul and value of ceramics lie in innovation and personality, which are also important standards in distinguishing between art and craft. In the long history of China, the ceramic art of Jingdezhen has walked away from creation to inheritance. While past periods produced different styles, such as the white porcelain of the Tang dynasty, the shadowy blue ware of the Song dynasty and the famille rose of the Qing dynasty, the artistic achievements of the majority of modern Jingdezhen ceramists rarely depart from tradition. One exception is the work of Linhong Li, professor of fine arts at Jingdezhen Ceramic Institute.

Professor Li places emphasis on the personality and creation of art, rather than common customs. Early on, he studied oil painting and wood cuts. Later, he received recognition

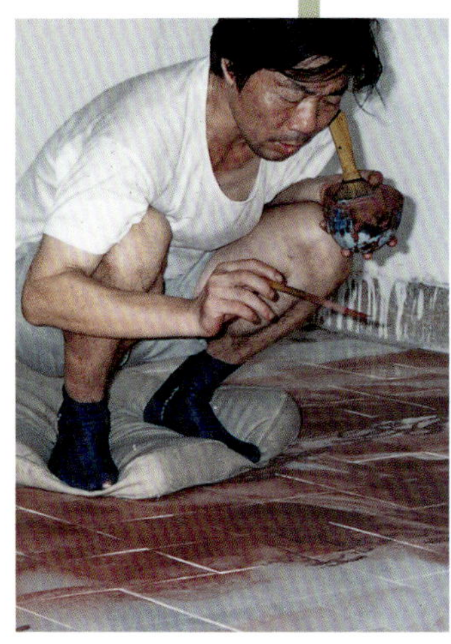
Pigments are applied in two stages: the first for negative or cool colors; the second for warm colors.

for his traditional wash painting. Yet he remained unsatisfied with his achievements, until he began working with glaze-painted slabs. There is an idiomatic saying in China: "Deep roots give rise to flourishing leaves." It was his foundation in ceramics combined with his knowledge of oil painting, wash painting and gouache that led to this innovative work.

The slab is formed by dry pressing. After a 2192°F firing, a white glaze is applied and the slab is fired again to 1832°–2012°F. On the fired glaze, Li brushes fluxed pigments in two stages. The first is for negative or cool colors, which are then fired to 1472°F. The second is for warm colors, subsequently fired to 1400°–1436°F.

Professor Li's ceramic painting style is regarded as "breaking through the constraint of traditional ceramics." While he places emphasis on materials, he uses modern painting techniques to aesthetic advantage, creating a mysterious, graceful realm of color.

Slab painting, approximately 24 inches square, with white glaze and fluxed pigments, multifired.

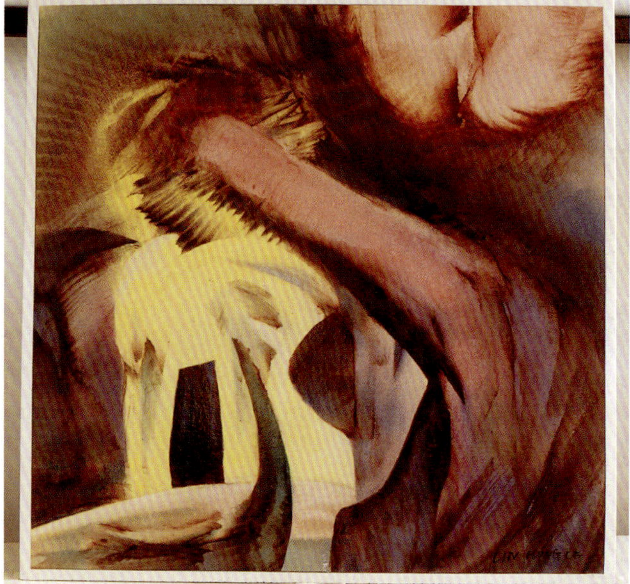
Multifired dry-pressed slab painting, approximately 24 inches square, by Linhong Li.

Thomas Orr
Ceramic Paintings

by Kate Bonansinga

The images that artist Thomas Orr renders on the back of his ceramic paintings are as important as those on the front. They are reserved for the collector or the brave and curious viewer who takes the time and initiative to lift the pieces, weighty as they are, off the wall and turn them over. These slab-built multifired earthenware "canvases" are about 3 inches deep. Many of them look like terra-cotta bricks in size and shape. Others are larger and squarer, referencing windows.

His work incorporates geometric shapes: squares, ovals and X's in bright colors: light blue, brilliant orange, yellow and green. Some pieces have images of house forms; others are landscapes. The houses and landscapes tend to be darker in color, and can be thought of as a continuation of Orr's interest in home and history. Conceptually, the weathered, unearthed quality of the front plays off the concealed painting on the back, just as the study of history involves digging up and decoding artifacts or unknown information.

PHOTOS: PHIL HARRIS

"Green Wall/Black Door," 14 inches in height, multifired earthenware.

Orr began working with clay in the early 1970s, but he didn't begin painting until he was in graduate school. There he was surrounded by painters and felt free to experiment with clay as a painting surface. He also painted on cinder blocks and suitcases. Thus began the sculptures for his "House Series," which developed into his graduate thesis installation, "Going Home."

"Five Windows," 25 inches in length, earthenware with slips and glazes.

The houses in "Going Home" and in ceramic paintings may refer to shelter and protection, but they may also refer to confinement or a false facade. While color in many of his graduate-school pieces was less than successful because they were "too rushed, the idea of color wasn't resolved," he overcame this problem by layering slip and glaze color.

Process

Orr's process of attaining his distinctive palette supports the content of his work: he creates a history in each of his pieces by layering glaze upon glaze, color upon color. He begins with a coat of white slip on bone-dry greenware, which acts as the equivalent of gesso on wood or canvas. He then bisque fires the pieces in an oxidizing electric kiln.

On the abstractions and houses, Orr applies several colors of glaze, each isolated from the other, and fires to cone 04, again in an electric kiln. After this first glaze firing, the surface has a glossy, wet quality. He then applies more glazes, perhaps the same ones as the first time, perhaps different ones, and fires a second time to cone 06.

For a dry, flaky appearance, he uses glazes with a high magnesium carbonate content. In certain places the glossy, initial layer of glaze shows through, which conveys a sense of age and depth. Orr then adds more glazes and fires at least a third time, often more, repeating the layering and firing as many times as necessary to achieve the desired surface texture and color. He sometimes completes a piece with a cold finish of wax, paint or polish.

His approach is somewhat different for the landscapes, which are new to his repertoire. If the abstractions and houses are mature, the landscapes are in their infancy. On these long, narrow, horizontal pieces, he

"Three Sons in the House," 18 inches in length, earthenware, with brushed slips and glazes, fired to cone 04 in an electric kiln, by Thomas Orr.

mixes the glazes directly on the surface, rather than applying each color separately in a discrete area.

This technique is reminiscent of a painter blending colors on canvas, an activity that may be the inspiration for the final sentence in Orr's artist's statement: "Sometimes I feel like a painter trapped in a potter's body."

"Three Sons in the House," synthesizes the concepts of home, landscape and abstraction. It is a wall box that refers to a window that offers access to the world beyond oneself. The background is pale green and is surrounded by a bold, black painted outline. Three orange spheres in the upper left corner float above a yellow elliptical shape balanced sideways on five black legs. To the right is a large, black rectangle within which is a hint of a pointed roof line. Orr intends for the viewer to "understand the work through his or her own unique insight," just as we understand our and others' histories, experiences and homes.

Recipes

Clay Body

Cone 04

Barium Carbonate	2 lbs.
Talc	8
Wollastonite	8
Ball Clay	25
Cedar Heights Redart	100
Fireclay	25
Grog (60 mesh)	16
Grog (30 mesh)	10
	194 lbs.

White Slip

Cone 04

Feldspar	3 parts
Ball Clay	1
EPK Kaolin	6
Silica	3
	13 parts

Apply to bone-dry greenware.

White Crawl Glaze

Cone 06

Borax	3.7%
Gerstley Borate	44.1
Magnesium Carbonate	29.4
Zircopax	5.1
Kaolin	17.7
	100.0%

Dry Matt Glaze

Cone 04

Barium Carbonate	9.3%
Gerstley Borate	18.7
Magnesium Carbonate	18.7
Nepheline Syenite	28.0
Kaolin	16.1
Silica	9.2
	100.0%

For color variations, add commercial stains.

Chartreuse Glaze

Cone 04

Gerstley Borate	5 %
Lithium Carbonate	80
Silica	15
	100 %
Add: Bentonite	2 %
Chrome Oxide	3 %
Tin Oxide	7 %

Lizard Glaze (Black)

Cone 06–04

Borax	6.3%
Gerstley Borate	43.8
Lithium Carbonate	9.4
Magnesium Carbonate	25.0
Nepheline Syenite	12.5
Silica	3.0
	100.0%
Add: Chrome Oxide	2.7%
Copper Carbonate	31.0%

Regina Heinz
Interactive Canvas

by Paul F. Dauer

"Polyphonic," 47 inches in length, stoneware, with oxides, stains, slips and lithium glaze, multifired.

Whether wall-hung reliefs or freestanding, the ceramics of Regina Heinz, an Austrian expatriate living in England, transcend convenient and conventional categorization as vessel or sculpture. Her sculptural forms integrate vessel components and her vessels are per se sculptural.

Initially trained as a painter, Heinz turned to ceramics after immigrating to England. She studied throwing with Takeshi Yasuda and coil building with Magdalene Odundo at Goldsmith's College in London, but gravitated to slab construction as a foundation for glaze and slip applications in the style of abstract painting.

Unlike those artists who use ceramic forms as surfaces on which to paint, she relies on the interaction of glaze, stains, clay body, texture and contour to create expression. As Heinz has noted, clay imparts its own unique characteristics; for example, the stress cracking from surface distortion during initial

"Homage to Paul Klee," 27 inches in height, stoneware.

slab preparation or from pushing or stretching the malleable slabs to produce volumetric forms.

Her current work encompasses wall-hung panels, sculptural stellae and pillow forms. The latter range from forms suggesting a new pillow, puffy with stuffed goose down, to tired, well-worn chair cushions irrevocably bearing the imprint of their occupants. Whether pertly puffed or deflated with age and use, these forms are enlivened by surfaces born of multiple firings of lithium glazes, slips, oxides and stains. The palette is dominated by vibrant electric blue, brick red, gray brown or off white, with an occasional yellow highlight.

Process

Heinz works in grogged stoneware, either White St. Thomas or Crank Clay. The latter is more gritty from grog, the former smoother but still gritty. The grit imparts greater texture to the surface and visually complements the speckling introduced by underfired lithium glaze. She notes that the grog also provides strength for control in the forming process.

Each form is fired initially to 2012°F in an electric kiln, with subsequent glaze firings at 1895°F in an oxidizing atmosphere to maintain a matt finish. At that initial temperature, the lithium glaze just begins melting, which creates depth and variation in the background coloration. Patterning is introduced through copper inlays, which produce black lines sketching her subject. Solid bands or slashes of color are the result of masking off areas of incised design. These are accented with colorful stains and oxides.

Her themes are abstracted geometric patterns, extracted and interpolated from photographic "sketches." These photo essays are intrinsically valued artistic études on their own, but also serve as a catalyst for further abstract expression when transferred to the clay surface.

Heinz's interest in photography began even before her earliest art studies. From the first, her pictures were intuitive abstractions, making

"Space Map," 13 inches in length, slab-built stoneware, with oxides, stains, slips and lithium glaze, multifired in an electric kiln.

sense from random compositional structure. The geometric patterning, which she translates from photo to clay as a structured abstraction, is simply another evolutionary step.

With its bold color strokes and defining lineation, her work correlates most to, but is clearly distinct from, that of Gordon Baldwin among ceramists, or Franz Kline and Paul Klee among painters. The latter is admittedly an inspiration; one of Heinz's works, a freestanding torso-like menhir, is entitled "Homage to Paul Klee."

Ironically, or perhaps logically, her wall reliefs were her earliest ceramic works, and transitional to more conventional ceramic forms, rather than, as might be surmised, an experiment emerging from ceramic traditionalism. Again, the painter's origins explain this evolution—clay becoming a three-dimensional canvas for painting with glazes.

The wall plaques grew into the pillow forms, conceptually back-to-back painting surfaces joined to form a continuous volumetric surface. These are constructed from slabs over a paper and wire armature. Heinz describes the process as one of conceptualizing the form, which is "upholstered" in clay much

"Breaking the Ice," 39 inches in height, constructed from soft stoneware slabs, bisqued, brushed with lithium glaze, oxides and stains, fired to 1895°F, by Regina Heinz.

as fabric might be tailored to a mannequin. In assembling these pillows, she is conscious of and articulates both inner and outer surfaces, distorting the slab from the obverse to create an undulating, sensuous outer surface.

Heinz also explores wall multiples, a series in which several plaques make up a whole work. Some consist merely of 6 to 12 related component plaques. Others may be physically related by connectors, aluminum rods that pierce and join groups of plaques. These explorations were prompted by a desire to compose works commanding a larger area.

The most successful of these multiples is perhaps appropriately entitled "Breaking the Ice." Its motivation was a photo of tufts of melting snow interspersed by patches of ground. Abstracted to six panels, each 6 by 8 inches, it captures the essence of this jigsaw of natural patterns.

Color and Form

by Judy Seckler

Lime Group, up to 6 inches in height, handbuilt with air-brushed underglazes.

Emily Rossheim's dialog with clay has been ongoing for more than 30 years. Largely self-taught, she's run a full-time, one-person ceramics studio, where the most recent fruits of her labor have been pristine, glowing bowls saturated in color. So striking are the colors that people have told her they look like they have lights inside of them.

She established a comfortable rhythm of producing wholesale work early in her career. As a result, her studio didn't rely on retail sales or supplemental income from teaching. But lately, Rossheim decided she needed a new creative challenge, and has been exploring photography and digital imaging.

Rossheim has charted new (yet once-familiar) territory in the studio-potter apprentice system. In order to do this, her new take on this is actually more like historical models based on the European guilds, where a master teaches an apprentice how to make the work of the master. Whereas many potters have made work until their last dying breath, Rossheim's signature designs are now being crafted by Tom Marrinson, an accomplished potter in his own right. Rossheim handles accounting, tracks customer contracts, pays the bills and distributes income, fills out applications to shows, and handles the firing schedule and delivery. In addition, she takes care of promotion, pays taxes and does the filing and invoicing. Eventually, Marrinson will take over the studio. When the transition is complete, Rossheim will have a whole new set of skills

Centerpiece, up to 4½ inches in height, commercial underglazes, fired to cone 04.

to begin another creative business while Marrinson assumes ownership of the studio with the benefit of an established client base.

"It was the right fit," says Cassandra Corcoran, a clay artist who worked in Rossheim's Vermont studio for three years. Corcoran discovered she was too social to live the life of a full-time potter and instead took on the role of ceramic matchmaker. She met Marrinson at the school their children attended and thought he would be just the right candidate to become Rossheim's partner and fulfill her desire to leave her studio in capable hands. "She's had a 'child' all these years, but now she's ready to let it go," explains Corcoran. "The child is moving into another home."

Rossheim's minimalist bowls vibrate with color. According to her, their genesis began 15 or 20 years ago when she wondered what kind of vessel would complement the succulent plants that she loved. After some experimentation, she created planters with inner containers to house the "desolate, alien plants." She says that the public's interest in them has never slowed down. Over time, her forms have become simpler and more graceful. "There's nothing to hide behind," she says.

Once the two potters found they were compatible, Rossheim began to pass along her knowledge of all aspects of the business. Marrinson learned her methods of production, how to organize wholesale shows,

Hot Bowls, up to 4 inches in height, layered underglaze, once fired to cone 04.

planning the year's work schedule, ordering raw materials, packing and shipping work and keeping organized records of all transactions in the business.

Process

Seldom at a loss for ideas, Marrinson looks forward to many productive years in the studio. "And when my time with the work is done," he says, "if someone is willing to take it over, hopefully I can pass it on with the same patient guidance that was bestowed upon me."

Rossheim has found that the color of her bowls speaks loudly to her audience, so both Rossheim and Marrinson add only the tiniest bit of texture to finish off a vessel.

Through trial and error, Rossheim discovered that a box of wooden pencils that she sharpened and taped together create an effective pencil stippler. Once pieces have dried to the right hardness, the stippler can be applied to the outside surface of a bowl to create the textures to add interest to the piece without overshadowing the form and color. The only hitch is to watch the amount of pressure being applied. Too much force punctures the surface and requires repair. If the work can't be repaired, it's back to the drawing board.

She and Marrinson low fire white earthenware at cone 04 to get pure clean color. Mostly, commercial underglazes are used, while some custom mixing is done. Many layers are

Cool Bowls, to 5 inches in height, white earthenware with applied texture.

Marrinson demonstrates the stippling technique that Rossheim perfected for the surface of her vibrant bowls.

applied to get a good application. The bowls are colored, inside and out, and fired once. The interior and exterior color combinations demonstrate a sophisticated and skillful use of color. This saturated color engages audiences on a more emotional, rather than intellectual, level.

Rossheim has taken her design cues from a clean, minimalist aesthetic. She was always reserved about working with color when it came to painting and drawing, but somehow she felt a certain freedom working with clay. She also draws inspiration from the work of studio potter Richard DeVore. Whereas his forms have a deceptive nonchalance, the Rossheim/Marrinson bowls exude formality.

Over the years, Rossheim has established a healthy client base at wholesale craft shows, and orders are scheduled 6–12 months into the future. Marrinson no longer frets about taking ownership of the studio. He's busy refining his skills and techniques to keep up with orders more easily. Rossheim has generously passed along her knowledge, while supporting Marrinson's ideas about how to put his own stamp on the work. While he acknowledges that the university setting was an amazing arena to expand his technical and aesthetic horizons, it was lacking in one aspect, "Once you have a vision of your work, what do you do with it?"

Porcelain Slip Glaze

by Joseph Godwin

During a summer spent in picturesque Switzerland, I dreamed of painting impressionist landscapes, sunlit fields of golden rape-seed flowers and chocolate-brown wooden houses surrounded by bright red geranium flower boxes. But it rained every day, all day. Instead, I painted psychological portraits of a wet landscape and read C. G. Jung books on psychology. My paintings emerged as abstract, inner landscapes—nonobject and nonrepresentational.

Painting in Switzerland was a welcome change from many years of porcelain carving at my studio in Massachusetts. While in Switzerland, I also visited several potteries around the country; at one of these, the potters were developing stoneware slip glazes and I gave them the recipes for the porcelain slips I had developed for slip carving and inlay. By the time I left Switzerland, they had developed a series of opaque, stoneware slip glaze colors.

On returning to the States, I decided to continue working with the painting techniques I had studied that summer, but using porcelain slip glazes on porcelain vessels. I began by formulating a slip glaze with the same basic flux that the Swiss potters had used for their stoneware slip glazes—wollastonite. A natural calcium silicate, it is used to replace silica and whiting in clay bodies and glazes. My base test consisted of a combination of wollastonite and bone-dry Grolleg porcelain, equal parts by weight. The original 50:50 clay and wollastonite recipe, which had produced a matt stoneware slip glaze for the Swiss potters, fired to a semiopaque porcelain slip glaze when mixed with a commercially available, cone 8–10, Grolleg porcelain body consisting of approximately 50% Grolleg kaolin, 25% G-200 feldspar and 25% silica, plus 2% Veegum T. I have found that 50 parts of the Grolleg porcelain body

"Narrow-Necked Vessel," 18 inches in height, wheel-thrown porcelain, with brushed and trailed slip glazes.

Porcelain vase, 18 inches in height, wheel thrown, brushed with porcelain slip glazes, fired to cone 8.

with 40 parts wollastonite fluxes as well as the 50:50 recipe.

Since so much of the slip glaze is a clay body, its original chemical and physical properties have a significant influence on the maturation temperature and application properties. A slip glaze formulated with a cone 8–10 porcelain body will have a lower maturation temperature than one made with a cone 9–11. A slip glaze with the proportions of 50 parts porcelain and 40 parts wollastonite will attain semi-opacity or translucency when fired to the same cone (or slightly lower) as the porcelain body. The addition of 8%–10% Zircopax to the basic slip glaze will create an opaque white slip glaze; 4%–5% will yield a semi-opaque slip glaze.

Lithium compounds in the form of petalite or spodumene (lithium feldspar) can be an important flux in porcelain slip glaze formulation. Lithium carbonate and lithium fluoride are also potential flux additives. They extend a slip glaze's firing range and in some cases help to control crazing in translucent slip glazes. Wollastonite itself has the property of reducing shrinkage in clay bodies and glazes, thereby preventing crazing problems in the opaque white slip glaze. When combined with Zircopax, it prevents crazing in the Opaque White Slip.

A lithium compound combines well with a frit of a low-fire feldspar, such as nepheline syenite, in a translucent slip glaze. Nepheline syenite contains a large percentage of soda and potassium in proportion to its alumina and silica content. This composition categorizes it as a low-fire soda spar. It fluxes well with a lithium compound, such as spodumene. Approximately 5% frit or nepheline syenite, combined with 5% petalite or spodumene in a slip

"Little Round Vessel," 5½ inches in height, with Cerulean Blue beneath Delft Blue, Golden Yellow, Orange and Red Porcelain Slip Glaze.

glaze, can expand the vitrification range of a cone 8–10 slip glaze to cone 6. These fluxes also afford stronger color saturation of colorant oxides and stains.

The basic purpose of a porcelain slip glaze is to facilitate the glazing of both green- and bisqueware with user-friendly versatility. Relatively little had been accomplished in this vein until the arrival of modern deflocculants and drying agents, which keep the liquid slip glaze in suspension for application purposes and allow it to dry correctly on bisque, thus preventing crawling during the glaze firing.

I prefer to use Veegum T for slip glaze suspension. It is a processed, colloidal magnesium alumina silicate that is used as a plasticizer in commercial porcelains. It consists of 80% Veegum, an inorganic bentonite, and 20% CMC gum, an organic binder. The colloidal property of 2% Veegum T significantly increases plasticity in porcelain clay bodies. The presence of 1% Veegum T creates an excellent deflocculant for porcelain slip glaze; it causes a mild thixotropic reaction. The addition of 1% bentonite increases thixotropy.

Frequent stirring during glazing with a porcelain slip glaze that includes 1% Veegum T is unnecessary. It takes several hours for a porcelain slip glaze containing Veegum T to complete a colloidal, mild thixotropic reaction, during which time a thin film of water forms on the surface as the slip glaze gels into a suspension rather than settles to the bottom of the bucket as glazes tend to do. When shaken or stirred, the slip glaze returns instantly to its former liquid state. It is therefore important to mix dry slip glaze recipes with a measured quantity of water, not only to ensure the correct consistency for your particular application purpose, but also to guarantee the correct consistency for a thixotropic glaze suspension.

If the slip glaze is mixed too thin, excess water will cause an uneven suspension, rendering the mixture unsuitable for glazing. The excess water will hold only the finer slip glaze particles in suspension, while most of the slip glaze forms a stiff mass at the bottom of the bucket. Excess water might not be decantable, without removing some of the finer slip glaze particles, for several weeks in a large volume of slip glaze.

As a safeguard, I measure the correct volume of water for a fairly thick slip glaze solution, then carefully thin the slip glaze to the desired consistency during and after sieving. The ratio of 1 cup of water per 10 ounces of dry slip glaze mix creates a consistency of slip glaze thick enough to brush and thin enough to pass easily through a 100-mesh sieve. Because a slight amount of water can significantly alter the viscosity of a small volume of liquid slip glaze, I thin a cup or two of thick slip glaze with squirts of water from an ear syringe.

A drying agent helps ensure even drying, which in turn prevents thick layers of slip glaze from lifting from the bisque surface during application. Stain colors containing metallic oxides tend to increase the surface tension of a slip glaze. This impedes adhesion, resulting in the drying slip glaze's tendency to crack and peel away from the bisque surface. Glycerine works well as a drying agent, especially for successive layers of slip glaze colors involving variable thicknesses applied onto moist bisque. The addition of 1 tablespoon of glycerine (6.25% fluid volume) per 1 cup of slip glaze is adequate.

Each time I dig into a pile of bone-dry porcelain shavings beneath the trimming wheel, I feel as if I am rediscovering the concept of glaze. To mix a porcelain slip glaze base, I add an equal weight of wollastonite to the bone-dry porcelain, or four parts wollastonite (by weight) to five parts bone-dry porcelain, depending upon which slip glaze base is desired. First, I dry mix outdoors, then pour the mixture into a container of steaming hot water premeasured by volume. The hot water quickly slakes the bone-dry scraps of porcelain, and the slip glaze is ready to pass through a 100-mesh screen within minutes. That's all there is to mixing a container of porcelain slip glaze from scraps, as the correct proportion of Veegum T is already included in the porcelain clay body.

I need only add small proportions of opacifiers or fluxes for translucency, then the colorant(s) of choice, and glycerine for application.

Either base slip glaze combines well with commercial stains, but it is important to test each color, because some commercial stain colors are elusive at high temperatures. To mix a color test, I add the percentages of oxides and stains for a particular color to 10 ounces of thick liquid slip glaze base (prepared by mixing 10 ounces of dry slip glaze with 1 cup water) and resieve.

The above ratio of water to dry slip glaze creates a thickened slip glaze ideal for brushwork on greenware and bisque. An additional 20% water (by volume) thins the glaze enough for dipping or pouring on bisque. Adding glycerine as a drying agent is unnecessary for dipping and pouring. I mix a large liquid volume of each slip glaze base, but add the glycerine only to small containers of colored slip glaze, after I mix in the colorants and resieve.

Porcelain slip glaze has become a process for me to transform clay and glaze into color expression. Since I have chosen to immerse myself in this glazing technique, color composition has become the major theme of my work. Wheel-thrown vessels have become objects to radiate color, and the forms have begun to emerge from the wheel more free flowing. With porcelain slip glaze, my work has gained momentum, developing a free-flowing connection between the greenware and bisque stages.

Porcelain bottle, 13 inches in height, with layered porcelain slip glazes, by Joseph Godwin.

Recipes

Opaque Slip Glaze Base

Cone 8–10

G-200 Feldspar	13.9%
Wollastonite	44.4
Grolleg Kaolin	27.8
Silica	13.9
	100.0%
Add: Zircopax	8.0%
Bentonite	1.0%
Veegum T	1.0%

The opaque recipe fires to a white gloss at cone 8. The addition of 8% Zircopax is optional when formulating opaque colors with some stains, such as reds and yellows, because they contain opacifiers; 4% Zircopax is often sufficient for opacity.

Semiopaque Slip Glaze Base

Cone 8–10

G-200 Feldspar	11.9%
Petalite (or Spodumene)	4.8
Wollastonite	47.6
Grolleg Kaolin	23.8
Silica	11.9
	100.0%
Add: Bentonite	1.0%
Veegum T	1.0%

The semiopaque base fires to a white semimatt on porcelain at cone 8, but will become translucent at cone 10.

Translucent Slip Glaze Base

Cone 8–10

Wollastonite	42.1%
Ferro Frit 3269 or Nepheline Syenite	5.2
G-200 Feldspar	13.2
Grolleg Kaolin	26.3
Silica	13.2
	100.0%
Add: Bentonite	1.0%
Veegum T	1.0%

The translucent slip glaze recipe utilizes frit and petalite additions to create translucency at cone 8. The presence of lithium helps to prevent pinholes from forming in slip glazes containing frit and stains.

Some frits and stains reach their maximum glaze-use temperature below cone 10. A translucent slip glaze containing these may begin to boil and form pinholes unless it contains a percentage of lithium to extend the recipe's maturation temperature range.

When mixed with trimming scraps from a cone 8–10 Grolleg porcelain body, the recipes are as follows:

Opaque Porcelain Slip Glaze

Cone 8–10

Wollastonite	4 oz
Zircopax	1
Bone-Dry Porcelain Body	5
	10 oz

Semiopaque Porcelain Slip Glaze

Cone 8–10

Petalite (or Spodumene)	0.5 oz
Wollastonite	5.0
Bone-Dry Porcelain Body	5.0
	10.5 oz

Translucent Porcelain Slip Glaze

Cone 8–10

Petalite (or Spodumene)	0.5 oz
Wollastonite	4.0
Ferro Frit 3269 or Nepheline Syenite	0.5
Bone-Dry Porcelain Body	5.0
	10.0 oz

Mix each of the above scrap-clay recipes with 1 cup hot water, then add 1 fluid tablespoon glycerine for brushwork on bisqueware.

Porcelain slip glazes formulated with oxide and stain additions, and fired in oxidation can achieve maximum color saturation. Muted color tones and textures can also be achieved by layering Volatile Black Slip Glaze beneath opaque white and colored slip glazes:

Volatile Black Slip Glaze

Cone 8

Hardwood Ash	4.2%
Wollastonite	18.7
Temmoku Glaze	52.1
Bone-Dry Porcelain Clay	25.0
	100.0%
Add: Black Stain	4.2%

Temmoku Glaze

Cone 8

Gerstley Borate	10.9%
Whiting	6.5
G-200 Feldspar	76.1
Kaolin	4.3
Silica	2.2
	100.0%
Add: Red Iron Oxide	8.7%

The volatile black melts at a slightly lower temperature and bleeds through the outer layers to create a mottled surface similar to reduction-fired stoneware in which iron particles in the stoneware clay speckle the surface. Varying proportions of oxides and stains have a significant effect on glaze melt and surface texture according to the flux capability of each colorant. Black colorants tend to have a strong fluxing quality.

Recipes

Black glossy and black matt slip glazes can be differentiated with a slight alteration in the ratio of whiting to silica in the form of wollastonite. The following example substitutes half of the wollastonite with whiting in the glossy black recipe to create a black matt:

Glossy Black Porcelain Slip Glaze

Cone 8–10

Wollastonite	42.9%
Bone-Dry Porcelain Body	57.1
	100.0%
Add: Zircopax	7.1%
Black Stain	7.1%

Matt Black Porcelain Slip Glaze

Cone 8–10

Whiting	21.4%
Wollastonite	21.4
Bone-Dry Porcelain Clay	57.2
	100.0%
Add: Zircopax	7.1%
Black Stain	7.1%

A satin matt slip glaze can be calculated by altering the ratio of whiting and flint found in the base recipe. To alter a translucent slip glaze containing a frit into a satin matt, a ratio of 30% whiting to 10% wollastonite is introduced. The following satin matt slip glaze fires to a satin texture that works especially well with red colorants:

Satin Matt Porcelain Slip Glaze

Cone 8

Petalite	5%
Whiting	30
Wollastonite	10
Ferro Frit 3269	5
Dry Porcelain Clay	50
	100%

Tests have resulted in a Super Opaque Porcelain Slip Glaze that gives excellent results over Delft Blue as well as Glossy Black:

Super Opaque Porcelain Slip Glaze

Cone 9

Wollastonite	33.33%
Zircopax	11.11
Bone-Dry Porcelain Body	55.56
	100.00%

For color variations, try adding up to 10% stain.

The original opaque recipe does not hold a strong white when applied over blue and black slip glazes. However, I continue to use it to layer between colored slip glazes and to glaze the insides of vessels.

A full palette of porcelain slip glaze colors is the most important part of my porcelain slip glazing process. Complex color combinations can be created by layering translucent colors over opaque and semiopaque colors so that they flow and pool. For the following color variations, add oxides and/or stains as specified to 10 (or 11) ounces of liquid porcelain slip glaze.

Ruby Red

Mason Stain 6001	6 grams
Mason Stain 6003	6 grams
Mason Stain 6006	6 grams
Mason Stain 6031	6 grams

Peach Bloom

Ferro Pink Stain	18 grams

Cinnabar Red

Ferro Pink Stain	12 grams
Reimbold & Strick Stain K2323*	12 grams

Orange

Ferro Pink Stain	6 grams
Reimbold & Strick Stain K2323	6 grams

Golden Yellow

Reimbold & Strick Stain K2323	25 grams

Yellow

Reimbold & Strick Stain K2323	12 grams

Aqua Green

Mason Stain 6201	3 grams
Mason Stain 6364	6 grams

Deep Green

Mason Stain 6202	3 grams
Mason Stain 6263	6 grams

Cerulean Blue

Cobalt Carbonate	0.25grams
Copper Carbonate	1.20 grams
Mason Stain 6364	18.00grams

Sky Blue

Mason Stain 6363	3 grams
Mason Stain 6364	6 grams

Turquoise Blue

Mason Stain 6390	25 grams

Cobalt Blue

Cobalt Carbonate	6 grams
Red Iron Oxide	3 grams

Delft Blue

Copper Carbonate	3 grams
Red Iron Oxide	1 gram

Violet

Mason Stain 6319	6 grams
Mason Stain 6385	6 grams
Reimbold & Strick Stain K2323	6 grams

Salts of the Earth

by Diane Chin Lui

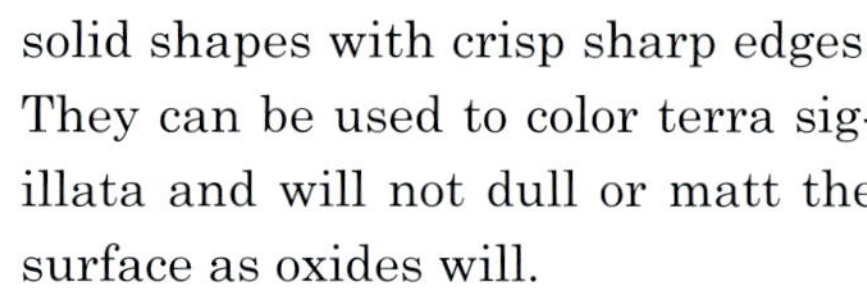

Latex resist was painted on the lip and underside of this porcelain vessel and 10% potassium dichromate was painted on the entire bowl. The latex was then removed and the following WSMS solutions were dotted and brushed on: 15% cobalt chloride, 50% cobalt chloride, 25% iron chloride, 50% nickel chloride and an "all gray" solution (10 grams each of potassium permangantate, cobalt chloride, molybdic acid and iron chloride in 100cc water).

Beautiful, soft, muted-color brushstrokes and washes of water-soluble metal salts decorate Gary Holt's translucent porcelain bowls and plates. The simplicity and quiet presence of his works belie the years that Holt spent experimenting and perfecting his technique. Using water-soluble metals salts (WSMS) demands excellent technical skills and careful attention to details.

Water-soluble metal salts are often compared to watercolors in application and decoration. They produce a variety of interesting effects on ceramic works, such as halos of color, fumed or smoky halos, solid shapes with soft, diffused edges or solid shapes with crisp sharp edges. They can be used to color terra sigillata and will not dull or matt the surface as oxides will.

CAUTION

Water soluble metal salts are extremely toxic and should always be used following the utmost safety precautions. Carefully read and adhere to the guidelines on the following pages whenever using these salts.

Holt has been testing and experimenting with metal salts for more than twenty years, while running a successful pottery studio in Berkeley, California. With little research literature available on WSMS, he has had to develop his own techniques through trial and error.

Chemistry

Water-soluble metal salts are simple solutions that are composed of nitrate, chloride and sulfate forms of metals, which dissolve in water. They are simpler solutions in comparison to glazes, which are usually composed of fluxes, alumina and silica, as well as oxides, carbonates or stains, and which may contain met-

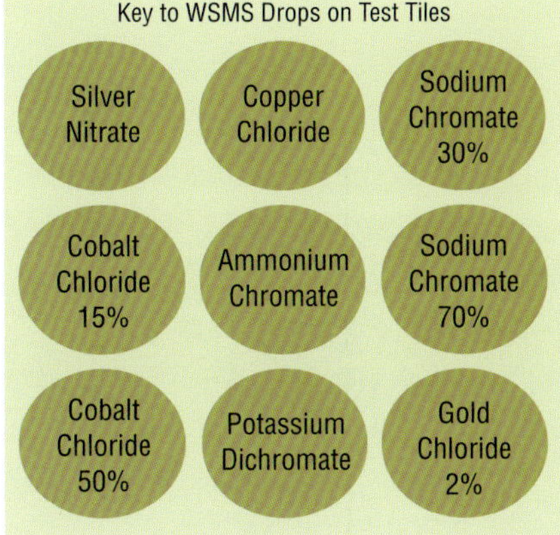

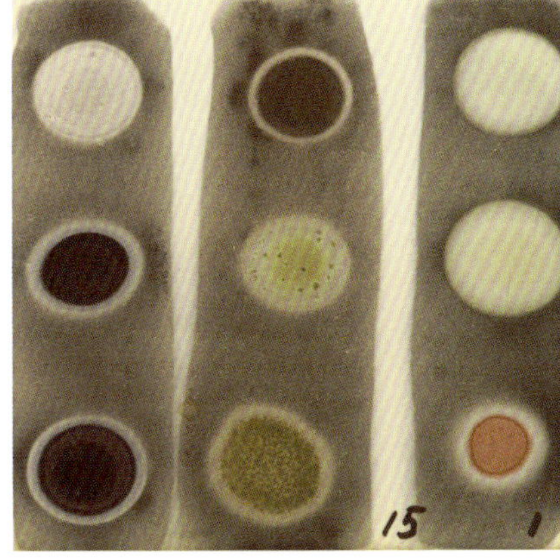

Background: Molybdic Acid

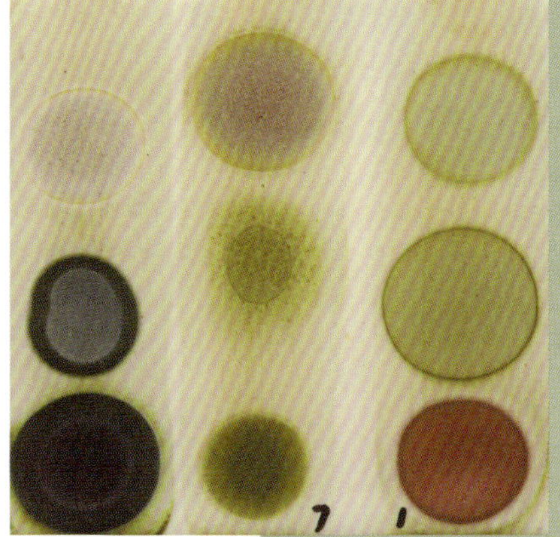

Background: Ammonium Chromate

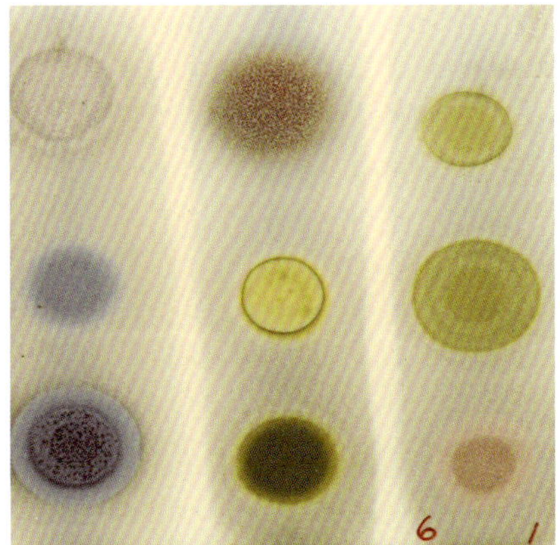

Background: Copper Chloride

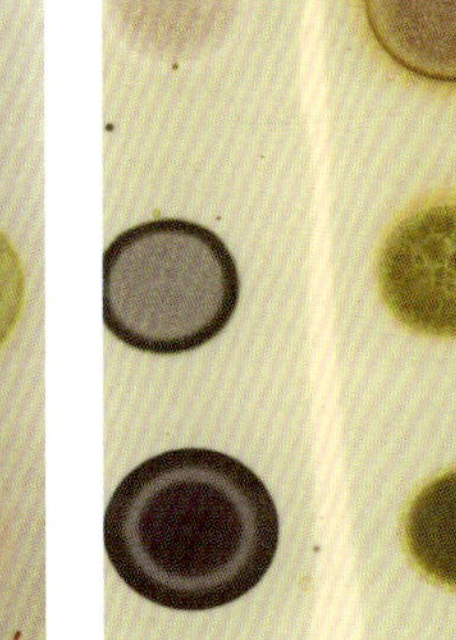

Background: Sodium Chromate 30%

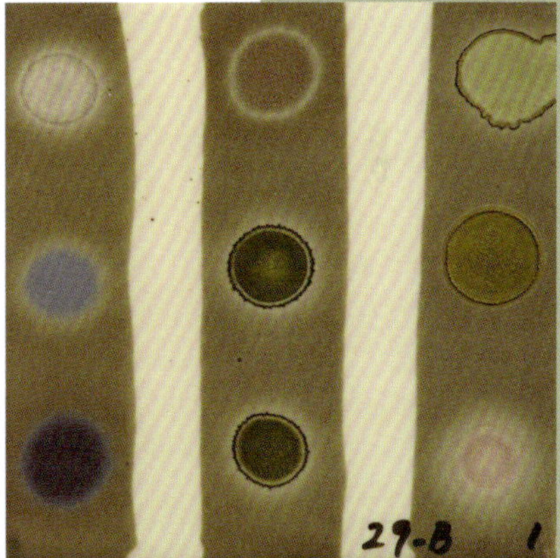

Background: Vanadyl Sulfate

al elements. Metal carbonates and oxides are the most commonly used form of metals in glaze, but more than twenty water-soluble metal salts may also be used (see chart).

Application

Holt prefers to use Southern Ice porcelain, formulated by Australian ceramist Les Blakebrough. The plasticity of the clay compares to Limoges porcelain clay. It does not have bone ash as part of its body. Holt likes Southern Ice for its translucency and whiteness, and he has noted that the color of the clay affects the brightness or clarity of the metal salts. The darker the color of the clay body, the more muted the colors will be. As an alternative to porcelain, Holt suggests a white clay body, stoneware covered with white slip or plain stoneware.

Often the color may "sink" into the clay body, which may or may not be desirable. Holt applies an opaque glaze to the inside of a pot if he does not want the color to migrate to the other side of the wall. Also, to keep his metal salts on the surface of the wall longer, he uses a nonreactive thick-

After latex resist was applied to the lip and underside of the bowl, a 2% gold chloride solution was painted around the entire bowl and a 50% tin chloride solution was painted in broad vertical strokes. The latex was removed and a 15% cobalt chloride solution was painted in dots and stripes on the inside and outside of the vessel.

COLOR	WSMS
gray	copper chloride (heavy application and heavy reduction can give pinks and reds) palladium chloride ruthenium chloride selenium (selenous acid, selenium toner) silver nitrate tellurium chloride vanadium (vanadyl sulfate, vanadium pentoxide)
blue	cobalt chloride molybdenum (molybdic acid)
green	ammonium chromate nickel chloride potassium dichromate sodium chromate
brown	iron chloride (iron chloride emits heat when mixed with water so the water should be added gradually in small amounts)
pink/ purple/ maroon	gold chloride (1–5% solution, adding either cobalt, manganese or tellurium will give different shades)
yellow	praseodymium chloride (very pale color)
black	cobalt chloride (50% solution) and iron chloride (100% solution) cobalt chloride (50% solution) and nickel chloride (50% solution) NOTE: neither of these combinations will yield a true black, just a close approximation.

ener. The thickener has the added effect of intensifying the colors.

Firing

Holt states that the clay vessel or form must be bisque fired between applications of metal salts. This technique is called "setting" the color. All water-soluble metal salt colors are temperature sensitive. The colors will change depending on the firing temperature.

Practical and Safety Concerns

It is absolutely essential to observe safety and health precautions when using these materials. Holt refers to the Merck Index whenever he uses an unfamiliar material. As always, the potential hazards depend on the concentration of the chemicals used and the safety practices of the ceramist. Holt believes everyone can use WSMS with the required attention and care. A Materials Safety Data Sheet (MSDS) should accompany each product when purchased. If the supplier does not provide an MSDS, buyers should ask for one. These information sheets will provide the precautions for storing, using and disposing of the products.

Water-soluble metal salts should be stored in containers separate from regular glaze mixtures. The containers should be well labeled to avoid any accidental mixing of the chemicals. In addition, acids and bases should be kept in separate containers.

A NIOSH-approved respirator should be worn when measuring

Latex resist was painted on the lip and underside of the bowl. A 15% cobalt chloride solution, a 50% cobalt chloride solution and a 50% tungsten solution (with a small amount of sodium hydroxide to help dissolve the salts) were applied with an eye dropper onto the surface of the bowl. Phosphoric acid was added with an eye dropper to create halos by "removing" the central area of a previously painted color.

Latex resist was used to mask two rectangular areas before applying salts. On the left, a 50% cobalt chloride solution was painted on. Then phosphoric acid was added to create dots. On the right, a 2% gold chloride solution was painted on. A 50% tin chloride solution was dotted on with a brush. A second bisque firing was done to set the colors. Then a 30% vanadyl sulfate solutionwas painted on the left and 15% iron chloride was painted on the right. Separate solutions of 15% cobalt chloride, 50% cobalt chloride and 50% nickel were dotted on with an eye dropper.

and working with water-soluble metal salts so the chemicals are not inhaled or ingested. Eye goggles should also be worn, especially when using acids. For hand protection, Holt wears two sets of gloves—a latex glove over a nitrile glove—because skin can easily absorb these chemicals.

Besides the health and safety concerns, local laws and regulations regarding the proper and safe disposal of the chemicals should be checked. When mixing the chemicals, Holt mixes only small amounts so that the disposal of the remaining solution is kept at a minimum.

Though the health and safety concerns are numerous and may appear overwhelming, they are necessary precautions to a rewarding and exciting facet of ceramics decoration that has been explored by few ceramists. Holt continues to experiment and add to his extensive body of knowledge on the subject and generously shares this knowledge through seminars and workshops. As evidenced by the fruits of Holt's experimentation, water-soluble metal salts present many possibilities for new forms of expression in ceramics.

WARNING

These materials are toxic. You must read and understand all safety precautions on the previous pages before using these materials.

Recipe

Water-Soluble Metal Salt Solutions

For 5% solution

Water-soluble metal salt	5	grams
Water	100	ml

For 10% solution

Water-soluble metal salt	10	grams
Water	100	ml

*For 15% solution**

Water-soluble metal salt	15	grams
Water	100	ml

As a rule of thumb
5% solution = light color
10%–15% solution = medium color
15% and above = intense color

Intensity of the color may be deepened by layering the color. However, most colors will not become darker once the surface is saturated with a 5% solution of the water-soluble metal salts.

*Potassium dichromate has a 12% maximum solution. More KCr^2 will not dissolve.

Equipment needed

- Respirator
- Chemical-resistant gloves
- Protective goggles
- Triple beam scale
- Beakers
- Graduated cylinders
- Foam or bristle brushes
- Bisqued pieces

Joyce Jablonski
Layered Surfaces with Decals

by Kathleen Desmond

"I am an artist and I pursue that way of being. I've always been an artist," proclaims Joyce Jablonski, professor of art and head of the ceramics program at Central Missouri State University. Her primary attitude about artmaking is a "quest," as she describes it, for understanding. She seeks to understand her physical, psychological and spiritual world through her artmaking. She looks into herself. She studies Jung's views on psychic and spiritual energy in the human psyche. Philosopher Ellen Dissanayake writes, "Artmaking is about looking into yourself and finding your humanity and finding out what makes things special." She says every culture finds a way to make things special, and this "specialness" becomes ritual and spiritual.

Fascinated by the dualities in nature, in mathematics and in words, Jablonski is thoroughly engaged by natural shapes and forms. She considers the psychological, the intellectual, the cultural and the spiritual. Her way of being is a ritual process of thinking and believing. She keeps a Joseph Campbell quote in her journal that is meaningful to her: "New metaphors emerge in a modern medium for the old universal truths." If her work could be categorized, it would be in terms of ritual art. Process as ritual. Ritual as content.

Ritual art has been an integral part of human experience throughout history. Ritual is inextricably connected to the extraordinary, to the sacred and to the very nature of human psyche. Prehistoric shamans understood the spiritual qualities of their environment. They drew on cave walls and conducted ceremonies to evoke spirits. Contemporary ritual artists make objects and tell stories that speak to our psyche and to our soul. Ritual artists bring new meaning to ideas and objects. Ritual art currently enjoys a contemporary convergence in the artwork of Joyce Jablonski.

Jablonski questions the traditional definition of the sacred and the secular. She considers herself a

"Catacomb #1," 60 inches in height, slab- and coil-built terra cotta, with slips and glazes, fired multiple times. The "Catacomb Series" refers to the organic architectonic quality of pod shapes as vessels.

modern-day shaman who examines, challenges and brings new meaning to the ordinary. Jablonski investigates her "self" through an intuitive drive for creating works of art and addressing the value of objects beyond appearance.

An overriding function of ritual is as an interface between concepts or systems and human beings. As such, ritual makes ideas and methods visible and accessible. Ritual also can function as an interface between human beings and the natural world. Nature is seen as a model of ultimate truth and rituals are used to both educate people about that order, as well as to bring human activity into alignment with it. Sometimes ritual emphasizes objects and experiences that are intended to be more than what they might appear to be. These ritual objects and experiences transcend their limited material qualities and become extraordinary.

Ritual art comes through the hands of artists who transcend themselves, or the work itself, in the process. Transcending is the capability of being out of control in a sense, or of being a product of an extremely compulsive, fixated temperament, thinks Dennis Oppenheim. When an artist is in such a psychological state it is possible for work to assume ritualistic proportions, partly because of the psychology of the artist and partly because of the kind of latent maneuvering of the art concept as it becomes disengaged from the artist. Ritual artists are highly charged and extremely engaged.

Ontological tension between spiritual and cultural attitudes is why Jablonski went to Ghana, Africa. Similar to when she left Ohio for graduate school in Texas years before, she found herself in a completely new environment. Everything was different; the trees, the landscape, the environment, the culture. "I wanted to figure out what the environment was all about, the cactus, the desert, the vegetation, the ocean." The history of artmaking ideas and processes with clay are in Africa for Jablonski.

She thinks of flowers not only as symbols, but as natural things in another context. The flowers are not just flowers or the formal flower shape, they are expressive and geometric, bilateral, biomorphic and anthropomorphic shapes. Meaning comes from a flower, a cowrie shell or the palm of the hand in Jablonski's work. She wants to add other organs to her organ series like brains and ovaries/pods. She wants to take the poetic intimate nature of these objects and put them in a different context that reinterprets their meaning (e.g. heart/passion/life, ovary/female/reproduction/sexuality).

"Object, notion, idea, material." Jablonski loves the poetic rhythms and patterns of words and music and colors and textures in visual art and in science. She likes to make up words for the pure joy of the rhythms and patterns. When she works with the rhythm or a pattern, or the patterning of a shape, it becomes more of a voice. It becomes the process of

"OV #1," 55 inches in height, handbuilt terra cotta, with slips, glazes, fired multiple times, plexiglass, steel and porcelain insulator base. Inspired by organs, Jablonski transforms the common toward the abstract.

making things, repeated images and patterns. "Artmaking is basically spiritual for me," says Jablonski.

Jablonski thinks music, mathematics, numbers, genetic code and art all have parallel meanings. She actively engages in creating patterns and rhythms and she believes in the repetition of random numbers to obtain individuality. Her influences are multitudinous. She delights in the large drawings she is making. She is engaged in the energy of each little mark. Each little mark can be added to another mark. She challenges herself to use different materials so she can find the genuine beauty in each material, in each new dimension. For a long time she only used black and white. "Line, shape and movement are a challenge with just black and white," she says. They become a pattern, a rhythm, a process, a ritual. She created stamps of flowers with their "ovaries showing" to incorporate in her formal graphite drawings. "The abstraction becomes primary," she explains. "It is not about making something representational. It's about good design, about how you walk through a drawing, how you enter and exit a drawing, and what attracts you."

Jablonski thinks of her drawings in terms of formal mark-making investigations. Both her printmaking and drawing have influenced a series of tiles made in Norway at the Porsgrund Porcelain Factory with artists Ole Lislerud and Suzanne Fagermo. Fagermo helped Jablonski design and pull 450 sheets of color-separated decals. Since then, she still investigates this process because she thinks her two-dimensional work influences her three-dimensional work and vice-versa. She found this process refreshing and considered it a new and challenging investigation.

I've heard Jablonski speak about the rhythm of the process of working with clay. She explains both practical and theoretical issues surrounding art making. She explains that it is important to control the shape of the clay. When she works with clay, it is as though she possesses it. She demonstrates such concentration and focus that it is an aesthetic experience just watching her. With such control and concentration she is able to lose control—to transcend—the clay, herself and any work she creates. Jablonski's ability to work with clay with such competence and passion, gives evidence to the quality of her attitude and her own work. Working with the process of artmaking with clay is as much the art as the objects she makes.

After a fire destroyed six years of her work, Jablonski engaged in the creative, spiritual and intellectual quest to create new work for her one-woman exhibition last winter at the Daum Museum of Contemporary Art in Sedalia, Missouri. All the work in the exhibition catalog was the documentation of work that no longer exists. The new work is the collective spirit of the ideas, images and studies that have been the core of Jablonski's spiritual, intellectual and psychological quests. The exhibition catalog documents the essence of her creative spirit. Clay has its own ideology and its own energy that was transformed and transcended by Jablonski after thoroughly engaging in the ritual process; the quest and rebirth of her art.

Decal Process

When Jablonski worked in the Porsgrund Porcelain Factory in Norway, she learned how to make decals and adhere them to clay tiles. She used the silk-screen printmaking process to make the decals. Instead of pulling inks across screens, as in traditional silk-screening, she pulled china paint across screens onto decal paper. She used five screens altogether, one screen of black images, three screens of color and one screen for lacquer. She printed the decals on 18×24-inch sheets of decal paper. Jablonski pulled the black and color separations first and then pulled a layer of lacquer to seal the china paints onto the decal paper. When applying the decal to the clay, the texture and surface of the decal depends on the texture and surface of the fired pieces of clay. For this reason, the process works best on glazed surfaces. If the decal is applied to a shiny and smooth surface the image will come out shiny and smooth.

PHOTOS: JERRY SCHMIDT

Tiles, 12 inches square, cast porcelain, with sprayed china paints over stencils and handprinted decals, fired multiple times, metal frame. Made in Norway with the help of Suzanne Fagermo and Ole Lislerud.

"Asexual Butterfly," 21 inches in length, slab- and coil-built terra cotta, with glazes, fired multiple times, by Joyce Jablonski.

Recipes

Red Terra Cotta Clay Body

Cones 06–02

Ball Clay	10.2 %
Cedar Heights Goldart	10.2
Cedar Heights Redart	71.5
Silica	8.1
	100.0%
Add: Grog	10.3%
Red Iron Oxide	2.1%
Bentonite	3.1%

Good for smaller work.

Base Slip

Cone 06

Soda Ash	25 %
Kaolin	75
	100 %

Add 15% stain of choice (more for reds).

Spray or use thinly or it will crack and peel.

June's Suede Base Glaze

Cone 06

Bone Ash	20 %
Gerstley Borate	80
	100 %

For dry suede finish, add desired stain (5–20% higher when using reds).

Velvet Crust Base Glaze

Cones 06–05

Alumina	33.4%
Gerstley Borate	49.9
Nepheline Syenite	16.7
	100.0%

Add desired stain 5–20% (higher for reds).

Add desired oxides for color 2–4%.

Organic Burnout Material

by Richard Burkett

"Oval Mesa 2," anagama-fired stoneware oval dish with ash glaze from the firing, with burnout texture at the bottom that includes small porcelain extrusions that were pushed into the clay.

In "Pressure Vessels," an ongoing series of ceramic works, I've tried to include visual references to both science and industry, as well as Midwestern life of the past century. Through the use of soda or wood firing and organic (carbon-based) materials included in the clay that burn out during firing, these vessels take on a corroded appearance. They represent failed vessels in their porosity, no longer being able to hold liquids or pressurization—castoff relics of another time. Many of the forms find resonance in memories of my frequent childhood visits to my father's chemistry laboratory, and my fascination with the many shapes of flasks, beakers, and other chemical glassware. Other works, more threatening in their references to weapons of war, address darker issues of the connections between science, industry, and the military industrial complex.

Pressure Vessel series, soda-fired stoneware with soybean burnout texture and found object lid.

Although the porosity left by the burnout materials is appropriate to the "Pressure Vessel" series, I have worked out ways to use these textures in other more functional forms that must hold liquids. I find that the burnout-material textures offer surfaces evocative of landscape and geological formations for an ongoing series of oval dishes that reference the Western U.S. landscape, especially mesas and buttes.

Process

Here's how to laminate clay containing burnout texture materials with untextured clay to allow more functional use as a serving container.

I've used a variety of grains like barley, dried soybeans, and buckwheat groats to create textures in clay. Other grains or dried seeds will work as well. Larger seeds like dried soybeans must be soaked first to allow them to swell before wedging them into the clay. If soybeans are added dry to clay and immediately thrown, they will absorb moisture from the clay, drying the clay somewhat quickly. As the soybeans swell in the drying clay, they expand and break the clay into small pieces. Similarly, seeds that are viable can sprout if kept moist, and the growing roots can break the clay. In figure 1 you can see a few vegetable burnout materials (clockwise from top): soybeans, buckwheat groats and barley.

The burnout materials need to be wedged into a small amount of clay (figure 2). The buckwheat groats shown here have been sprinkled across a thick slab that will then be rolled up and wedged. If you slice the wedged clay, you'll see the buckwheat groats added (left) and plain clay (right) (figure 3).

Clay with organic materials added must be used immediately to prevent rotting or mold. To create a functional form, you need regular clay for a liner. Begin by throwing

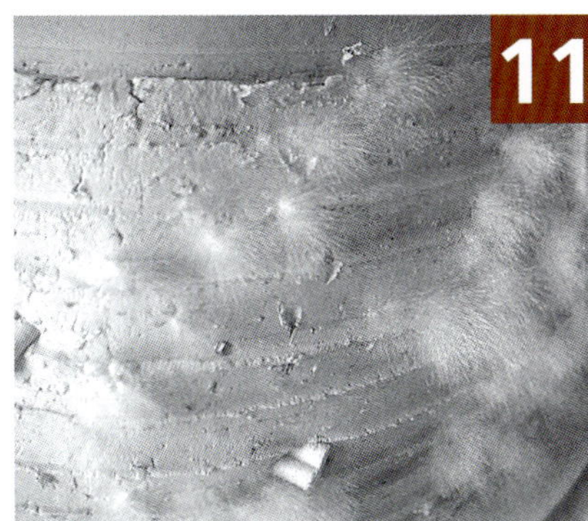

a thick, open-bottomed cylinder using burnout clay, then cut it off the wheel head (figure 4). Next, throw a smaller cylinder from regular plain clay and drop the textured-clay cylinder over the outside and roughly center it (figure 5). Join the two cylinders by expanding the inner cylinder against the outer texture clay cylinder, starting at the bottom and working slowly upward to avoid trapping air between the two layers as they are joined (figure 6).

Expand the two joined cylinders into the final shape. Here the top edge of the form has been cut with a wire to create a softly moving line (figure 7). Stretch the cylinder into an oval by pushing out the ends with your hand on the inside (figure 8). (Optional) After the form is pulled into an oval, push small extrusions of porcelain into the clay (figure 9). These extrusions can be made ahead of time and dried. Dip the dry extrusions in water right before impaling them into the clay wall to help them adhere.

After the thrown oval has stiffened somewhat, place it on a clay slab (figure 10), wrap in plastic, and allow to dry slowly overnight while the slab and oval reach the same moisture content. Organic burnout materials are best used and dried quickly (usually within a day or two), before mold like this (figure 11) can form. Beans rot quickly in wet clay!

Invert the leather-hard form onto foam (figure 12) to avoid damaging

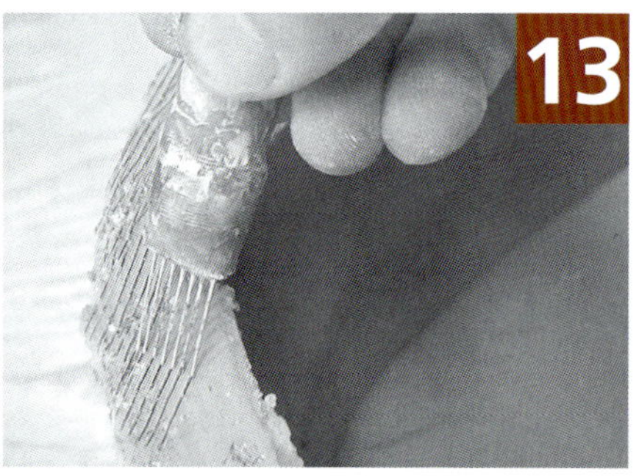
13

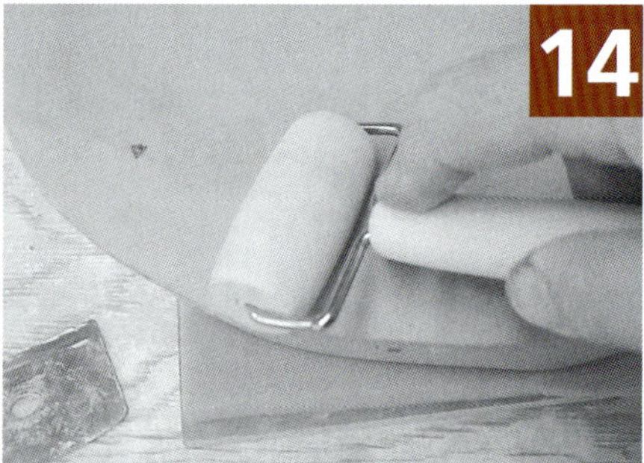
14

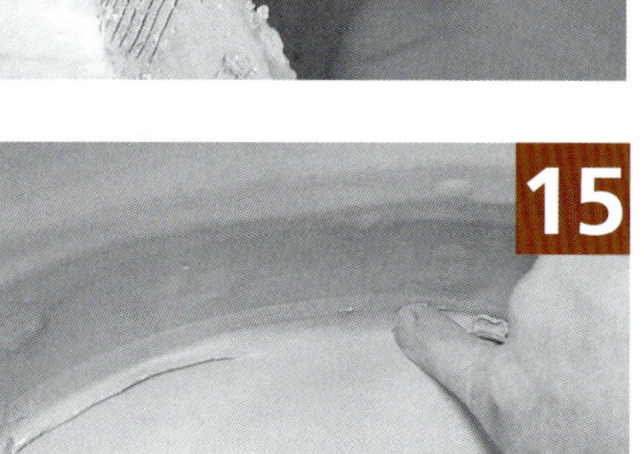
15

16

17

18

19

the rim while attaching the bottom. Score the clay deeply with a needle scoring tool to ensure that the bottom joint does not crack. Add slip and score again for a strong bond. Place the bottom on the oval after both joints have been scored. Roll and rib the joint to mechanically join the two parts (figure 14). After sealing the bottom from below, flip the form right-side up, then seal the seam on the inside with your fingertip (figure 15). Tip: A flanged lip has been left on the inside of the thrown form to add strength to the joint. Trim away the excess clay from the slab base (figure 16).

Run the pointed end of a wooden potter's knife under the slab base edge to help force the joint together and create a beveled undercut (figure 17). Additional trimming and smoothing can be done with a fettling knife and a rib. Flip the piece over using two boards and a piece of foam on the rim to avoid damaging the textured sides or distorting the slightly soft clay (figure 18). The completed oval "Mesa" dish, with small handles added to the ends (figure 19).

Firing Burnout Materials

Clay with burnout texture materials requires a special bisque firing. The bisque firing must be slow enough to allow the organic materials to burn out slowly. Especially important are both a longer-than-normal preheat soak of the kiln below the boiling point of water to remove extra moisture in the organic material used, and a prolonged temperature hold at about 400-420°F, just below the ignition point of carbon-based materials. Obviously, good ventilation is critical, both to burn out the texture materials completely and to safely remove gasses from the kiln (burning organic materials will produce carbon dioxide, or, without enough oxygen in the kiln, carbon monoxide). Gas kilns are preferred for bisqueing burnout materials, but bisque firing can be done in electric kilns with good ventilation equipment installed.

The following is a suggested firing schedule for burning out texture materials. Thicker sections of clay, or larger amounts of burnout materials, may require slower firing. (Thanks to Louis Katz for his suggestions on firing rates when I first started firing these pieces.)

1. Soak the kiln at 180°F for 4-12 hours, depending on clay thickness and the amount and size of texture materials, to remove moisture from the clay.
2. Heat the kiln slowly (20-50°F per hour) to about 420°F, and soak at this temperature for another 4-12 hours (or more if the work is quite thick) to allow the organic materials to carbonize and release as many gasses as possible. This is a very critical phase in the firing.
3. Continue the firing, with a similarly slow heating rate (20-100°F per hour, depending on clay thickness) to at least 600-700°F to allow the gradual burnout of the rest of the carbon-based materials. For thicker sections of clay, a soak for an hour or two at this temperature is recommended. Keep the kiln well ventilated until nearly red heat (1000°F) to ensure enough oxygen to burn out the organic texture materials.
4. Once the kiln has reached red heat and is past quartz inversion, the firing can proceed at fairly normal rates (100-250°F per hour), depending on the size and thickness of the work.
5. Normal cooling of the bisque is okay.

Tips for Success

- One challenge with organic texture materials in clay is to work quickly. Organic materials mold and rot quickly in the damp clay. Wedge the materials in the clay then finish working the clay within a day and let it start drying to avoid mold. (This might be a challenge in warm, humid areas!).
- Don't add scraps of organic texture material clay to your scrap barrel as it decomposes into a smelly mess.
- Burnout materials accidentally included in scrap clay that is reused for regular production can also cause blowouts in a bisque firing of normal speed.
- Chunks of wood or sawdust can also be used for other types of texture.
- Be careful not to use materials like plastics that can create extremely toxic gasses when they burn, or seeds that have been treated with antifungal chemicals.

Detail of anagama-fired burnout texture, small addition of barley along with coarse feldspar chunks.

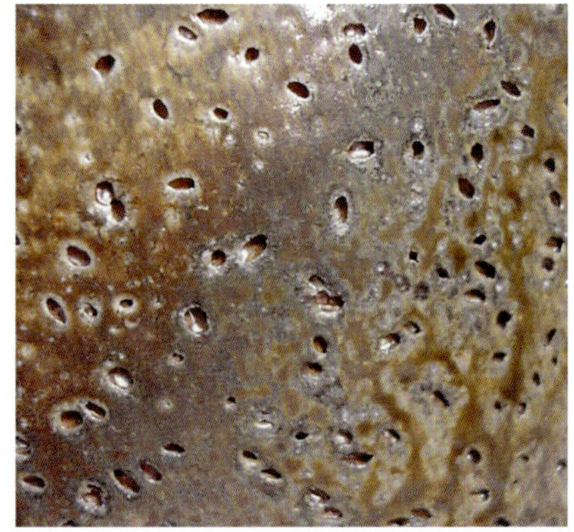

Anagama-fired burnout texture with a large amount of barley added.

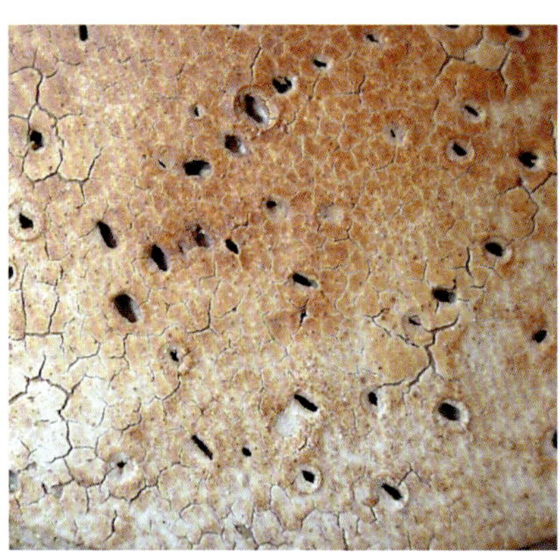

Crackle slip added over the burnout texture at the bisque stage. The form was then soda fired.

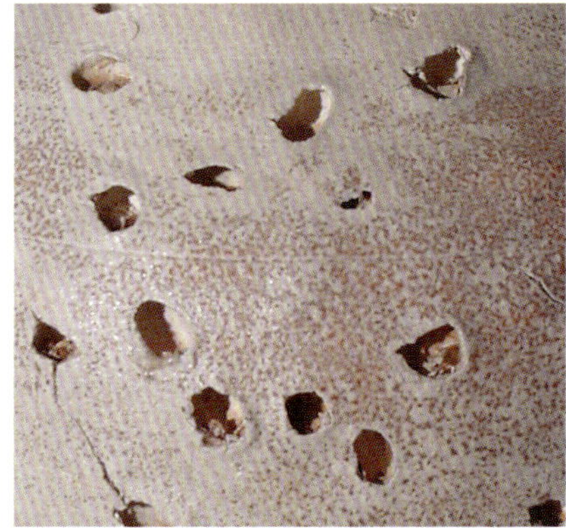

Soybeans create a much larger hole when they burn out. Soda fired to about cone 8 after flashing slip applied to bisque.

Buckwheat groats burnout texture with a very light soda vapor glaze.

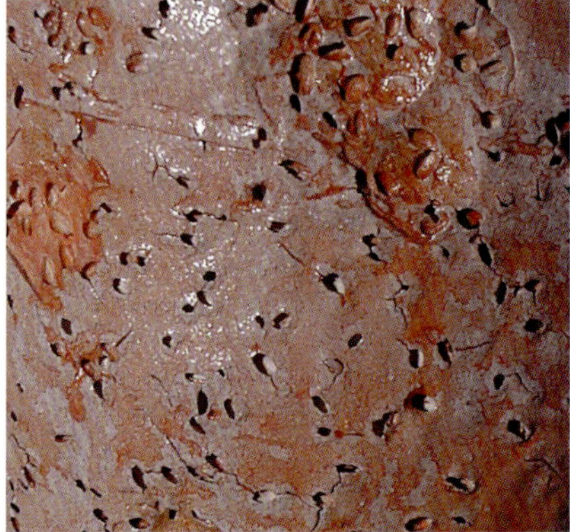

Bisque firing too quickly causes steam and burnout gasses to blow pieces out of the side of the form, or even completely destroy the clay form. You must bisque fire much slower than to allow the texture material to fully burn out.

Recipes

I prefer smooth light-colored clays for their contrast with the burnout texture. The following is a clay body I have used recently in soda firings.

III Clay (Version B)

Cone 10–11

Ingredient	%
Lincoln 60 Fireclay	25 %
EPK Kaolin	45
Kentucky OM-4 Ball Clay	15
Custer feldspar	10
Silica	5
	100 %
Add: Fine grog (or 60m sand)	6 %

I often use the following slips on ware to give an orange-red color in soda firing.

Fake Avery Flashing Slip #5

Cone 9–11

Ingredient	%
Nepheline syenite	23.8%
EPK Kaolin	43.6
Calcined Kaolin	29.6
Newman Red Clay	3.0
	100.0%

Opaque red orange brown fired in soda/salt kiln. Works well on bisqueware if applied very thinly (if too thickly applied it may peel). Increase the red clay by a percentage or two for darker flashing. Other red earthenware clays can be substituted.

Fake Avery Slip '02

Cone 9–11

Ingredient	%
Imco 800 clay	5 %
McNamee Kaolin	70
Nepheline Syenite	25
	100 %

This slip is best applied to greenware, and gives a nice cream to orange flashing in soda firing.

Pressure Vessel series. Soda-fired porcelain with soybean burnout texture and found object lids and steel shelf.

Tray, soda-fired stoneware with soybean burnout texture, with found object handle.

Pressure Vessel series, stoneware with soda vapor glaze, refired to cone 06 with terra sigillata, and steel lid and bale.

Vase, soda-fired stoneware with soybean burnout texture.

Pressure Vessel series, soda- and wood-fired stoneware, with found objects.

Amy Lemaire
Glass As Glaze

by Elizabeth Reichert

A group of bead pods, 2 inches in diameter each, soda-fired stoneware, with lampworked, soda-lime glass accents.

Artist Amy Lemaire wanders weekly through a local flower wholesaler hoping to stumble upon what she calls "oddities of nature"—a spiky pincushion protea, a droopy sandersonia blossom or an unexpectedly angled branch—that may eventually inspire her clay and glass beadwork. Like a scientist questioning natural patterns, Lemaire often wonders why a certain shape doesn't occur, and it is this that she sets out to create.

"I make things that do not appear in nature, but that I wish would," she explains. "I want these things to look like they grew out of the earth, as if a new species of plant pods were being encountered. I want the viewer to think: 'Is that a bead or a pod? What's its use? Was it found on the ground? Was it made in her studio, and if so, out of what?'"

An abstract painter by education, a bead artist by profession and a sometime floral designer by fancy, Lemaire and her work resist categorization. Unlike many of today's beadmakers, who work mostly within the ornamental traditions of the craft, stringing up their creations to adorn others, Lemaire's approach is more diversified. She wants to draw as much attention as possible to her beads (not the wearer) by making them large in a necklace, by mounting them as sculptural objects, or more radically, by using them to anchor submerged blossoms in what might traditionally be called a floral arrangement, but in her case certainly begs of another name. Such departures into the domains of floral design and sculpture embody that category-defying spirit—what many call thinking outside the box—that marks the way Lemaire works.

It is no wonder then that Lemaire began to question not only how a bead could be used, but also how a bead could be made. That was in Autumn 2004, months after she had set up shop in Chicago's Lillstreet Art Center, predominantly a clay studio. Surrounded by kilns and clay-covered neighbors, she began to notice that, while precious metal clay was commonly used by contemporary beadmakers, high-fire stoneware was as foreign a material to her colleagues as basket weaving reeds are to the traditional potter. After researching the surface compatibility issues between clay and glass, and the possibilities of constructing clay beads, Lemaire began to conduct experiments. Six months later, she discovered how to fuse clay and glass, a technique that, as far as this writer's research can conclude, is not being practiced by any other artist.

From a distance, the fusion of clay and glass seems a logical alchemy, one that, despite reason, has no place in history or in the contemporary craft genres. Some historians believe, for example, that glass originated as the accidental by-product of metal-smelting processes, others believe it was the accidental by-product of glazes being used without clay. After all, glass is, at its most basic level, a mixture of silicates—that same material which gives ceramic glazes their glassy shine. The only illogical component of fusing glass to clay arises from the difficulties involved in maintaining the temperatures of both the clay and the glass bodies so that cracking and separation do not occur.

Having worked with premanufactured beads since she was a child, and more seriously since she began making her own, Lemaire considers herself a "second-generation" participant in the Contemporary Glass Bead Movement. This movement is linked to the International Society of Glass Beadmakers, and supports juried exhibitions and publications meant to foster a more sophisticated study of the craft. While most within the movement draw their inspiration from the Murano and Venetian glass-bead industries, many use glass toward more contemporary ends by sculpting abstract shapes, or by encasing enamel-engraved landscapes within the bead. The most common technique used among these artists is called lampworking.

It was in 1997 that Lemaire was turned on to this technique after taking a workshop with Cindy Jenkins, one of the key players in the revival movement. Prior to this initiation, Lemaire's training had been self-taught and secret. She had strung together homemade polymer and

"Bead Pod," 4 inches in height, soda-fired stoneware bead, with lampworked, soda-lime glass accents, sterling silver stand. Lemaire prefers to present some of her beads on stands in order to draw attention to their sculptural qualities.

bought beads for her friends while following a more high-profile path as an abstract painter at the prestigious Art Institute of Chicago. The workshop changed her artistic focus: "I couldn't believe I could create something that would still be around in a thousand years. That day I was given the power and responsibility to create future artifacts. I came home with a torch and started buying books. I've been making beads ever since, and my painting has since shifted to the back burner."

Were it not for the lampworking method Lemaire might not have ever bonded clay and glass. The technique, again, widely practiced by today's beadmakers, involves a stainless steel rod, called a mandrel, around which a pencil-thin stick of glass is wound. The mandrel is covered with a clay slip (otherwise known as the bead release). The glass is heated by a table-mounted torch and does not stick to the steel as it melts because, as the mandrel is heated, this slip turns to powder, allowing the glass to be released from the rod. Lemaire's initial instinct about clay and glass fusion was peaked by that release agent. If the glass would adhere to the slip, she wondered, why wouldn't it stick to clay?

Lemaire began making clay beads then, all extruded and hollow, with the sides pushed in to resemble plant pods. Although she tried using porcelain beads, which were sensitive to thermal shock and separation during half of her attempts, and unglazed low-fire clay beads, which didn't even take to the glass, Lemaire eventually found her most consistent success with beads made from a high-alumina stoneware.

Because of both aesthetic preference and technical ease, Lemaire has worked predominately with the dot formation. She explains that the dot patterns are inspired by the color theories of Joseph Albers, a painter famous for claiming that "a color has many faces." His work, influenced by Geometric Abstraction and the Minimalism of the Bauhaus group, and consisting of consecutive tri-colored squares, greatly affected Lemaire during her painting days. Her beads thus retain this influence by maintaining a three-color balance. And yet in technical terms, Lemaire has had most success with the dots because the point of adhesion, and therefore, the point of possible fissure, is relatively small.

Wanting to eventually make larger, nonfunctional clay and glass sculpture, Lemaire is currently experimenting with sheets of glass and alternate clay bodies, while teaching her fusion technique at Lillstreet. Having trouble with thermal shock and separation when using larger surface quantities of glass, she has tried borosilicate glass, which has a lesser contraction and expansion rate. Likewise, she is building an insulation chamber that will protect the clay bodies, and hopefully lead to greater surface fusion by not allowing the ambient heat to be pulled away from the torch.

Despite the aesthetic and technical success of her beads—as floral design, as jewelry, even as curious, miniature sculpture that bridges a craft divide in a remarkably unprecedented way—Lemaire remains restless about how her work fits into not just the craft world, but also into "our American culture at large." When asked about the function of the bead in today's society, as well as the function of her recent discoveries, Lemaire initially brings up the fact that beads were not always used for self-adornment. In the Philippines they were used in marriage ceremonies. In ancient Asia they were scattered like seeds to induce plentiful harvests. And in North America, Europeans exchanged their beads as currency for beaver pelts.

"Traveling and studying these histories helped me realize the social significance of the bead in many cultures, including my own," she explains. "I now think that merging techniques and functions to create something new and progressive is identifiably American."

Merging techniques and functions is, of course, the precise motivation that has enabled Lemaire to create a body of work unique among that of her bead-making contemporaries. It will be interesting to see to what ends she will take her work, technically and aesthetically, in the lapidary or sculptural arts. "There is one thing about the craft world that does not fit with my personality," she explains. "And that is most artists who are working in glass work in glass, and most artists who are working in clay work in clay. There are very few who straddle the borders. And I am excited to finally be making work that is uncontainable in this sense."

Only time will tell, of course, whether her realization of clay and glass fusion will mark a landmark technical achievement in the chronicles of bead, glass and ceramic form. But in the meantime, hopefully other artists—clay, glass and bead practitioners alike—will follow Lemaire's lead, inspired to either dabble in other mediums, or adapt her clay and glass fusion technique to meet their own artistic ends.

"Woodland Pod Necklace," 8 inches in length from bead to bead, each bead approximately 1½ inches in diameter, soda-fired stoneware, with lamp-worked, soda-lime glass accents, Ghana seed pods, and leather.

PHOTOS: CINDY TRIM

"Single Pod," 2 inches in diameter, soda-fired stoneware bead, with lampworked, soda-lime glass accents, by Amy Lemaire.

Process

Lemaire uses high-alumina stoneware for her beads. The high alumina content causes more soda-glaze build-up, which in turn leads to more successful glass adhesion. The glass she uses is a soda-lime glass, the same used in Murano. However, the technical prowess behind Lemaire's beads does not occur only at the material level; it also occurs within the firings. The first two are those familiar to ceramists: a bisque firing followed by a soda-glaze firing. At Cone 10, soda ash, whiting and wood chips are added to the high-fire kiln in order to give the beads an "unpredictable and natural finish." What follows is the fusion firing. In a small test kiln, the clay body is heated to Cone 014, approximately 1540°F, the temperature at which glass begins to flow. Once the clay bead is red hot, Lemaire takes it out of the kiln with glass blower tongs. Using an oxygen/propane torch, she then melts the glass onto the clay body in a dot formation.

This fusion firing is followed by an annealing process familiar to glass and bead artists: the bead is placed in an annealing kiln and held at 968°F—a temperature relative to the type of glass used—for approximately 45 minutes. Then the kiln and bead cool down over the course of six to eight hours.

Gillian Parke
Feldspar Inclusions

by Kathy Norcross Watts

"Cherry Bomb," 12½ inches in length, wheel-thrown and assembled porcelain with feldspar/molochite inclusions, underglaze patina and inlay, celadon glaze, fired to cone 10 in gas-reduction; luster overglaze, open-stock decals, multiple firings to cone 017 electric.

North Carolina potter Gillian Parke's passion for finding harmony in what might appear contradictory is evident in both the pots she crafts and the life she leads. "The foundation of my work is contrast," she explains. She seeks it in surfaces, in images and in connections between Eastern and Western cultures.

Parke was born in Northern Ireland, but grew up in Weymouth, Massachusetts. Childhood visits to Ireland developed her appreciation for ceramics. Her grandmother would take her and her sister to a gift shop and buy the girls a piece of porcelain or crystal. "You couldn't touch anything," Parke recalls. "You could just look." Her introduction to the hands-on aspect of the craft occurred during the 1980s, when her mother owned a paint-your-own ceramics store called Gazebo Ceramics. Parke grew up amid molds, bisque ware, paints and lusters.

Despite her early exposure to ceramics, she never formally studied art in college, instead earning a B.A. in chemistry from Boston University. She did become involved in the arts; after a class lecture by a representative of the Museum of Fine Arts, Boston explained how his chemistry background helped in paper restoration, she decided to take a job helping to restore wallpaper. This experience piqued her interest in further study, so, in 1995, while working as an organic chemist for Glaxo Wellcome in North Carolina, she applied to graduate school.

Though she was accepted into graduate school in London, she received no financial assistance and deferred the degree for a year, then met her future husband and decided to stay in North Carolina.

"Wildflower Portal," 13½ inches in height, wheel-thrown and assembled porcelain with feldspar inclusions, underglaze patina, celadon glaze, fired to cone 10 in reduction; luster overglaze, open-stock decals, multiple firings to cone 017 electric.

After deciding against graduate school, while still working full-time, Parke signed up for her first pottery classes through the Chapel Hill Parks and Recreation Department. In 1998, she began working at Stone Crow Pottery in Pittsboro, making production tea bowls and cups for the owner. For her work, she was given use of the equipment. The task of duplicating forms, along with her work making glazes, helped her to gain technical skill.

She took various workshops and courses but found that she was just emulating what she'd been taught. "I didn't have my own voice," she said. She'd heard before the wisdom, "Don't borrow someone else's idea; steal it," which requires taking ownership of a technique in your own way. This advice helped her to develop her own style. "I've 'stolen' a lot of things from a lot of people and put them together to make what I want to say," she explains.

Brad Tucker, a resident potter of Cedar Creek Gallery for many years, advanced Parke's thinking further when she took a course he taught at Claymakers in Durham. "No one had really taught me about form and negative space," she says. "He would bring in all these different examples from his collection at Cedar Creek, and we would sit down and talk about why a pot worked." This was her first exposure to what makes a really good pot, she says. "He just had a profound effect on me, on the way I looked at pots."

From Michael Simon, whose work

was rough and more organic compared to Tucker's, she learned, "It takes quite a bit of skill to be organic." A potter must have the technical ability to create an organic form that works.

Others who have influenced Parke include the late master potter Sid Oakley, whom she met on one of her trips to Cedar Creek. Parke remembers that Oakley told her, "'My biggest advice is not to sell your work too early.' [and] I think what he said was true. I've never sold seconds. If a pot is not perfect, I just take a hammer to it. I feel like my pots are a reflection of who I am, and people are investing in my work."

When Parke was laid off from Glaxo in 2001, she decided to try working in pottery full time. "I found out that I couldn't be a production potter," she confesses, because she became bored with the repetition required.

She entered graduate school at East Carolina University in 2004 and gave herself a year to determine if changing careers was the right thing to do. During that time she gathered information from many sources. She gleaned wisdom from several other potters, all of whom left their mark on the way she would eventually create her pots. From Malcolm Davis, she learned about shino glazes and carbon trapping, while Chuck Chamberlain introduced her to lusters. From a workshop with Rimas VisGirda at the University of Indianapolis, she learned how color and line can affect one another

"Orange Daisy," Blue Rose 8½ inches in height, wheel-thrown and assembled porcelain with feldspar/molochite inclusions, underglaze patina and inlay, celadon glaze, cone 10 gas reduction; lusters, overglazes, open-stock decals, multiple firings to cone 017 in electric.

and gained more knowledge about painting with lusters. She also took Rimas' decal making workshop and created her own "bra-lady" decal, an image she still uses today. While researching to write a history paper on teapots, she learned about sexism as it related to the history of pots, and she became interested in feminism in art.

In 2005, she returned to Durham to begin working on her own ideas; she combined porcelain with feldspar inclusions, applied decals and lusters to create one-of-a-kind pieces. "I was trying to take the English porcelain tradition—white and feminine—and merge it with this Japanese Shigaraki tradition; it's very masculine," she explains of the process in which organic forms are fired with wood, and the flame hits the pot leaving nature to play a vital role in the outcome. As she adds layers of glazes and decals, each piece undergoes multiple firings to achieve the effects she seeks, sometimes as many as five to ten firings per piece. She wants her work to comment on the world. "Taking manufactured images and putting them on handmade pots is fascinating to me," Parke explains, adding, "I am interested in the conflict created by kitsch images on handmade objects, and in challenging the aesthetics and values presented when using such materials unconventionally. The resulting works illustrate the contrasts in aesthetics, forms, traditions and function found between Japanese pottery and fine porcelain."

Orange & Blue Daisy Teacups, 4 inches in height, porcelain with feldspar/molochite inclusions, underglaze patina and inlay, fired to cone 10 in reduction; luster overglazes, open-stock decals, multiple firings to cone 017 electric.

Throwing with Feldspar Inclusions

by Gillian Parke

I usually work in stages on a set of pieces using Highwater Clay's Helios porcelain with coarse Custer feldspar (1–10 mesh, Seattle Pottery Supply) and 50–80 mesh molochite wedged in. Throwing with the inclusions requires using a substantial amount of water to provide sufficient slip for lubrication. This helps prevent both finger cuts and tears in the turning clay. However, the piece will lose its strength and collapse due to the low plasticity of porcelain if too much water is used.

Each stage is thrown on the wheel and allowed to dry. Before removing the piece from the wheel, the feldspar and molochite matrix is exposed with a metal rib or trimming tool. This also serves to remove the surface slip.

After assembling the piece, it is completely dried and wax resist is painted onto areas that will eventually be glazed. Underglaze is then applied to the unwaxed clay areas. The underglaze is removed from the surface with a damp sponge, leaving an underglaze patina that accentuates the feldspar and throwing lines.

Wax resist is again applied to the dry surface. Using a needle tool, lines are etched through the wax, revealing the clay below. After wiping clean with a damp sponge, black underglaze is applied to the inlaid line.

After bisque firing to cone 07, wax resist is applied to black inlay lines so that glaze will not cover the line and affect the color. Glazes are applied by pouring, dipping and/or brushing. The resulting piece is then fired in a gas kiln to cone 10 in reduction.

Feldspar inclusions result in pearl-like eruptions covering the surface of the vessel. This surface is painted with various luster overglazes and fired in an electric kiln to cone 017 multiple times per layer of surface treatment.

Elaine Parks
Perfect Perforation

by Kris Vagner

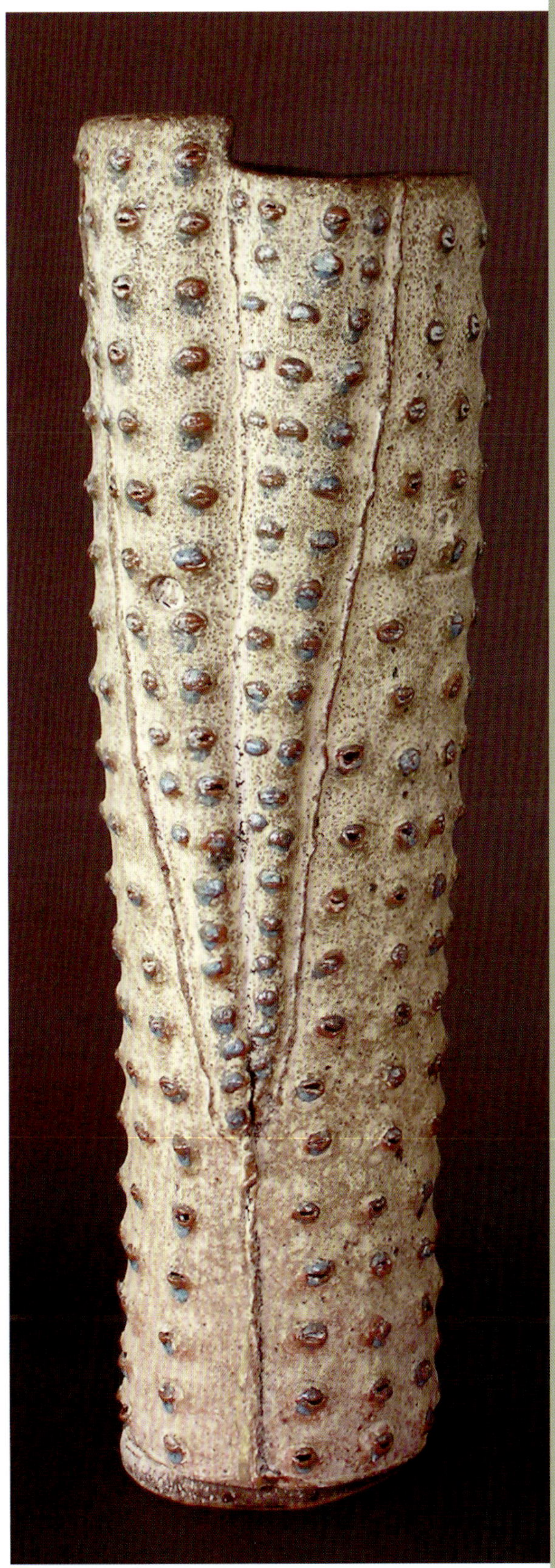

"Full Moon Canyon," 24 inches in height, earthenware, punctured, shaped and assembled slabs, cone 04.

"I'm trying to recreate the feeling I get from being in the landscape," says Elaine Parks. She's an artist in Tuscarora, Nevada, a town so small that any resident can tick off an accurate census count. Currently it's thirteen.

Parks' ceramic sculptures aren't shaped like the rugged, dry terrain or decorated with the purple lupine that carpets the hills in spring. But if you joined her on one of the hikes that punctuate her daily routine—from any house in town, you can walk to a few thousand square miles of open wilderness—you'd see a lot of the same textures and shapes you see in her studio.

Disembodied brick chimneys, crumbling stone walls and untended patches of poppies or rhubarb now decorate the sparse, gravelly lots where homes and businesses used to be. The town was built by miners and entrepreneurs during a gold rush in the 1860s and 1870s. In its heyday, it had a few thousand residents, but by the 1960s, when artists from urban areas started trickling in and out, the gold-rush population

"Canyon II," 16 inches in length, earthenware, punctured, shaped and assembled slabs, cone 04.

was long gone. Most of the original houses have succumbed to a century of heavy snows and dry summers. Some have been lying in splintery heaps so long they've become part of the scenery. Long-abandoned mining equipment, rusted halfway to oblivion, has littered the hills for so many decades that it seems more like part of the landscape than trash. Human industry and natural entropic processes have been competing for so long, the boundaries between nature and culture are sometimes blurred.

Parks' sculptures present a similar kind of overlap. Nature and culture both inform her aesthetic, which balances a primitive roughness with the polish of conceptual art.

A series of low-fire earthenware sculptures, some long like boats, some tall like vases or branchless trees, are influenced by the austerity of modernist sculpture and the organized chaos of the natural world. Parks explains, "I think of these pieces as something of a translation of the sensation of living in a remote place like this."

Parks' references to nature acknowledge its aesthetic and philosophical complexity. Some of her glaze effects—speckled matt grays, mottled greens with a waxy gloss—

resemble the surfaces of naturally worn rock or metal. Some look like the artfully weathered patinas of the car bodies that have been lying around town for decades, attracting photographers and plein-air painters.

Parks experiments with poking holes through the clay, puncturing the slabs in a grid pattern that covers most of their surface area. She says the puncture marks are inspired by the shapes she sees repeated on different scales in the landscape. Tiny pores in a rock and holes in the earth, which has been mined ambitiously around Tuscarora, are both comparable in shape to the holes in the sculptures.

Even though the sculptures' graphic elements are traceable to specific influences, they always stay within the realm of abstraction. Parks says, "I like the open-ended quality of working with an abstract approach. It gives viewers a direct experience without sidetracking them into ultra-technique. It more easily allows the viewer to make their own associations. I like to walk that line."

Wherever Parks' references to technique do become overt, they appear to allude to different mediums altogether. Some of the seams and darts from her sculptures look as if they're borrowed from welding or dressmaking. "There's a long history in ceramics of making clay look like other things, which it's good at," she notes.

"Green Pool," 9 inches in height, earthenware, punctured, shaped and assembled slabs, cone 04.

Usually it comes right back to the landscape. "I'm interested in the theoretical quality of the terrain, simultaneously on a grand scale and an intimate scale," Parks says. "When I see things in the landscape that remind me of my work, it makes me feel that I'm on the right track."

"Chinese Landscape," 18 inches in height, earthenware, punctured, shaped and assembled slabs, cone 04, by Elaine Parks.

Perfect Perforation

by Elaine Parks

1. I start by rolling out two or more slabs between ¼ and ¾ of an inch thick on a canvas-covered board.
2. Since I'm making tall forms, I cut out long, narrow rectangles for the cylinder wall.
3. I flip them onto a thick piece of foam.
4. I poke holes into the inside surface using either my finger or a small wooden tool, depending on what size hole I'm looking for. Sometimes, I draw on the front surface with a pencil while the slab is still on the canvas board, then transfer the slab to the foam and push out around the drawn lines.
5. I bend the individual pieces around forms, so they will set up in a curve. I usually use rolled towels and cardboard tubes from rolls of newsprint.
6. After the pieces get to a soft-leather-hard stage, I stand them upright and join them together. I don't let them get too set up, because I want them to be soft enough to push from the inside when the piece is together. Sometimes this part is a little tricky, getting the cylinder to stand up and get it joined while it's a bit soft, but I can get a more organic result this way. The curve of the individual piece is helpful. At this stage, I wish I had three hands.
7. To finish, I push the seams together to get them joined well, then I push out and in to get the texture how I want it.
8. After it sets up to firm leather hard, I lay the form on its side on the foam and beef up the seams with coils.
9. Last, I add the foot, which is quite thick. I do this when the form is upright first. When the foot sets up enough, I put it back on the foam and push the middle up to form a foot ring. There's a little back and forth—upright and laying on the foam—to finish the foot.
10. I dry the piece very slowly and then fire to cone 04.
11. I glaze using a combination of studio-mixed and commercial low-fire glazes. Some are painted on, and some are layered using a mouth sprayer. The sprayed-on glazes are mostly layers of very matt glaze.
12. Last, I fire again to cone 06.

Teruyama & Kelleher
A Collaboration

by Katey Schultz

In collaboration, it is always difficult to decipher where the work of one artist stops and another artist begins. Perhaps this is why the most successful collaborations speak in a new voice, a voice discovered spontaneously through joint exploration and the dissolution of ego. During the course of their three-year residency at Penland School of Crafts, Penland, North Carolina, artists and life partners Shoko Teruyama and Matt Kelleher have discovered how the concepts of ego and identity can dissipate through collaboration. In the process, their individual bodies of work have matured and their faith in the poetics of collaboration has flourished.

Originally from Japan, Teruyama began her formal studies in clay in the mid-nineties at the University of Nebraska-Lincoln (UNL), later earning an M.F.A. from Wichita State. While tradition and ceremony were a part of her daily life in Japan, it wasn't until leaving her homeland that she began to explore these concepts in clay. In developing her own body of work, Teruyama has created forms and surface designs that reference ceremony but contain a sense of humor and playfulness unique to her own vision.

Teruyama makes boxes, intimate bowls, small plates, vases and a variety of serving pieces. The work begins with bisque molds, slab construction and coil building to make thick, heavy forms. "I like to touch every surface when I'm working," says Teruyama, who carves, shaves and sands excess clay away to slowly reveal the final shape. Puffy handles, rippled or petal-like edges, and intricate patterns mark Teruyama's work as her own, which, in the end, captures a fine balance between calmness and celebrated intricacy.

"It was a big adventure to leave my culture behind. Now I look back at it and draw from it. I almost had to leave it to discover who I am, but that wasn't ever my intention. I love my country and my family, but there is a sense of freedom from everything

Large Jar with Bird Handles, 19 inches in height, wheel thrown with handbuilt handles and sgraffito decoration, earthenware, built by Kelleher and decorated by Teruyama.

Shoko Teruyama's Flower Plate with Bird Walking, 9 inches in diameter, slab built on a bisque mold with sgraffito decoration, earthenware.

when you leave something like that. The birds that appear in my work represent this sense of freedom," says Teruyama.

Sometimes the birds appear to dance, float or fly through Teruyama's signature vine and floral patterns. Other times, they adorn the edges of her work in various poses, such as her trademark owl smoking a pipe or a walking bird that wears Western boots. "This is my way of being playful," says Teruyama. "Birds are approachable. For me, I look at the owl

Large Bird with Lotus Necklace, 11 inches in height, slab built on bisque mold, coiled additions with sgraffito decoration, earthenware, built by Kelleher and decorated by Teruyama.

and think, 'What else would an owl be doing?' Owls are leisurely. They sit. Birds, they can go anywhere. They walk and move, so of course they're wearing Western boots. In my mind, it makes sense."

While Teruyama relies more on patterns and images to create mood, Kelleher's work seems almost minimalist in contrast. Mood is created through depth and color revealed in the soda-firing process, and he creates forms with every intention of allowing for this possibility. As Kelleher has expressed, "I combine a subtle balance of geometry in form, a comparison of symmetry and asymmetry in decoration and a serene surface. Softly, the work asks for the viewer's attention."

While he focuses on utilitarian objects for their universality, it is important to him not to be limited by process. Kelleher, who also studied at UNL, spent a long time searching for personal forms, perhaps the most notable of which are his trenchers. Trenchers, which look like robust dough bowls harkening back to the pioneers as they crossed America, allow for a maximum sense of depth with the slips and firing.

Kelleher uses pouring and layering techniques and applies minimal glazes over the slip to achieve a particular effect. "I want these forms to be like a window for the display into a vast landscape," he says. Much of his work also incorporates a single, bold, blue dot that punctuates the surface activity and creates immediate depth. At times the dot feels loud and close, other times it feels subtle and distant, as though resting on a far-off horizon. "That dot could be a bird, for example," says Kelleher. "What I like about the lack of specific meaning in a dot is that it can become more metaphorical."

By necessity, the cultivation of an individual body of work requires paying allegiance to some elements of craft while giving less importance to others. When considering his own work, Kelleher is so wedded to minimal use of slips and glazes that his forms act primarily as a vehicle for the expression of mood. His parameters for form are distinct, but the possibilities in soda firing are wide open. Teruyama, on the other hand, is so inclined to pattern and movement through imagery and lines, that her forms tend to be enjoyed more on display than in day-to-day use. Her surface design parameters are meticulous but the expression in her work is expansive. For both, the benefit of collaboration is release from some of these parameters.

Birds, as it turns out, have become the figure and form that most wholly embodies the new voice of the artists' collaborative work. While they also collaborate on large jars, small cups and various serving dishes, their latest and most provocative works are the large bird forms (roughly 12×24×13 inches), which are handbuilt by Kelleher and decorated by Teruyama.

"I kept joking with Shoko, telling her to just make the bird rather than spend so much time drawing it. Then I was working on lids for my fry pan forms, which worked their way into the large bird form. Now, when I work on the collaborative pieces, I feel very freed by the process, because I can work with familiar forms in an entirely new way, wondering how Shoko will decorate a certain piece," says Kelleher.

"The way you make work from the construction to the decoration stage is like a story from beginning to end," says Teruyama. "When I do my own work, I get to tell that whole story. With the collaboration, I have some idea of where it's going but it's never quite the same. The collaboration can be limiting, but it's also an interesting way to change myself. I enjoy the problem solving part of it. It's like, 'Hmmm...What am I going to do with this one?'"

Through collaboration, Kelleher has found freedom in form—an interesting switch considering that the primary message in Teruyama's individual work is that of freedom. Likewise, Teruyama has found a way toward self-discovery within the set parameters of an unfamiliar form—equally interesting given that Kelleher's individual work makes its personal mark first and foremost through form.

Finding a Third Voice

*by Matt Kelleher
and Shoko Teruyama*

Kelleher: The whole idea for our collaboration was to come up with something we could share the labor in and market as a third body of work. We set some parameters first: I wouldn't use any of my personal forms and Shoko wouldn't use any of her personal decorative motifs. We completed a number of pieces before admitting the lack of inspiration we felt from the work. I think our egos prevented the collaboration from growing.

Teruyama: At first, the collaborative work didn't look personal. We felt like we weren't in the work at all…and it's funny, because of course we knew that from our own experience. We knew that in order to feel good about it we'd have to be in it, but we just got so wrapped up in protecting our individualities that we forgot that lesson. The reason we ignored it is because we had so much ownership over our own ideas.

Kelleher: Our biggest growth during our three-year residency at Penland has been shedding those voices of what it 'should be' and finding voices of what 'could be.' Our idealism has since digressed, if that's the right word. We've grown into the point with our collaborative work where ego is almost gone. We have major ego with our own work still [laughter], but as far as our collaborations, we want to focus on making something that has both of us in it.

Matt Kelleher's Oval Trencher, 12 inches in diameter, slab built on a bisque mold, soda fired stoneware.